AF701935

POCKET BOOK GUIDES FOR QUANT INTERVIEWS

FE PRESS

New York

Pocket Book Guides for Quant Interviews

1. 150 Most Frequently Asked Questions on Quant Interviews, Third Edition, by Dan Stefanica, Radoš Radoičić, and Tai-Ho Wang. FE Press, 2024

2. Probability and Stochastic Calculus Quant Interview Questions, by Ivan Matić, Radoš Radoičić, and Dan Stefanica. FE Press, 2021

3. Challenging Brainteasers for Interviews, by Radoš Radoičić, Ivan Matić, and Dan Stefanica. FE Press, 2023

Other Titles from FE Press

1. A Primer for the Mathematics of Financial Engineering, Second Edition, by Dan Stefanica. FE Press, 2011

2. Solutions Manual – A Primer for the Mathematics of Financial Engineering, Second Edition, by Dan Stefanica. FE Press, 2011

3. Numerical Linear Algebra Primer for Financial Engineering, by Dan Stefanica. FE Press, 2014

4. Solutions Manual – Numerical Linear Algebra Primer for Financial Engineering, by Dan Stefanica. FE Press, 2014

5. Elements of Stochastic Processes: A Computational Approach, by C. Douglas Howard. FE Press, 2017

6. The Metropolis Algorithm: Theory and Examples, by C. Douglas Howard. FE Press, 2024

FE PRESS
New York

www.fepress.org

This edition first published 2024

Printed in the United States of America

ISBN-13 978-1-7345312-4-4

150 MOST FREQUENTLY ASKED QUESTIONS ON QUANT INTERVIEWS

Third Edition

DAN STEFANICA
RADOŠ RADOIČIĆ
TAI-HO WANG

Baruch College
City University of New York

FE Press
New York

To our beloved families

Contents

Preface to the Third Edition

The use of quantitative methods and programming skills in all areas of finance, from trading to risk management, has grown tremendously and accelerated with the advent of big data, machine learning, and artificial intelligence.

A core body of knowledge is required for successfully interviewing for a quant role. The challenge lies in the fact that this knowledge encompasses finance, programming, and several areas of mathematics and statistics. Moreover, brainteasers are often asked to probe the ingenuity of candidates.

The third edition of this best–selling quant interview questions book contains over 200 questions and includes for the first time statistics and machine learning questions, as well as new questions in other areas.

These questions are not only frequently, but also currently asked on interviews for quantitative positions, and cover a vast spectrum, from data structures to finance, statistics, stochastic calculus, and brainteasers.

The answers to all of the questions are included in the book and are written in the same very practical vein that was used to select the questions: they are complete and straight to the point – as they would be given in an interview.

Topics:

- Mathematics, calculus, differential equations.
- Covariance and correlation matrices. Linear algebra.
- Financial instruments: options, bonds, swaps, forwards, futures.
- C++, algorithms, data structures.
- Statistics and machine learning.
- Monte Carlo simulations. Numerical methods.
- Probability. Stochastic calculus.
- Brainteasers.

The authors are faculty members of the Baruch College Financial Engineering Masters Program, and have over 20 years of experience educating students who are very successful interviewing for quant roles. As such, the authors had the privilege to interact with generations of exceptional students, whose contributions as alumni to the continued success of our students has been tremendous. This book is a tribute to our special Baruch MFE community.

The *Pocket Book Guides for Quant Interviews* Series, includes two more books, *Stochastic Calculus and Probability Quant Interview Questions*, published in 2021, and *Challenging Brainteasers for Quant Interviews*, published in 2023. We hope that you will enjoy this trilogy and make best use of the books.

New York, 2024

Acknowledgments

As professors in the Baruch MFE program, it has been a privilege to have the opportunity to contribute to the early career success of so many talented students and alumni. The strong community that developed around the Baruch MFE program is the reward educators truly dream of, and, for us, a reality that is an inspiration. This book would have not been possible without our Baruch MFE community and we are grateful to everyone who is part of it.

Working alongside and learning from our colleagues, both academics and finance professionals, created the perfect context for writing this book; we are thankful to all of them.

Alejandro Cañete and David Zhang are owed a special thank you for their wonderful help on the programming questions, as is Max Rumyantsev who provided the art for the book cover of the first edition. Several students spearheaded the proofreading effort, and their help was greatly appreciated: Aneesh Subramanya, Gabriele Bernardino, Zhuo Cui, Niccolo Fabbri, Tengxiao Fan, Yiyang Fan, Ziyuan Li, Yuanting Li, Yicheng Sun, Yunze Sun, Suyang Yuan for the third edition, Zhaofeng Brent Liao, Sheng Rick Cao, Zilong Cheng, Xiangtian Forest Deng, Xiang Lu, Wei Mao, Bo Pang, Zhou Robert Qi for the second edition, and Yu Gan, Jun Hua, Svetlana Hagenbuch, Alireza

Kashef, Yi Bill Lu, Fubo Shi, Yujia Helen Sun, Jun Charlotte Wang, Peng Wu, Yanzhu Wendy Wu, Yongyi Ivan Ye, He Hillary Zhao, Wenyi Zhou for the first edition. It is a great pleasure to acknowledge so many former students who are now remarkably successful practitioners, and, much more importantly, wonderful people!

The tremendous support and the understanding of our families that made this book possible, and we are forever in debt to them. Thank you, thank you, and thank you!

This book is dedicated to all our loved ones, from the depths of our hearts.

Dan Stefanica

Radoš Radoičić

Tai-Ho Wang

New York, 2024

Chapter 1

First Look: Fifteen Questions.

1. Nine months call options with strikes 20 and 25 on a non–dividend–paying underlying asset with spot price \$22 are trading for \$5.50 and \$1, respectively. Can you find an arbitrage?

2. (i) What is the sum of the eigenvalues of the correlation matrix of n random variables?

   (ii) Find a lower bound for the sum of the eigenvalues of the inverse of a nonsingular correlation matrix of n random variables.

3. Let W_t be a Wiener process, and let

$$X_t = \int_0^t W_\tau d\tau.$$

   What is the distribution of X_t? Is X_t a martingale?

4. An 8×8 matrix contains zeros and ones. You may repeatedly choose any 3×3 or 4×4 block and flip all bits in the block (that is, convert zeros to ones,

and ones to zeros). Can you always modify the original matrix into an all–zero matrix using these block flips?

5. Find all the values of ρ such that

$$\begin{pmatrix} 1 & 0.6 & -0.3 \\ 0.6 & 1 & \rho \\ -0.3 & \rho & 1 \end{pmatrix}$$

is a correlation matrix.

6. Given a sample of size 1 from the normal distribution with mean μ and variance σ^2, with both μ and σ are unknown, give a *finite* confidence interval for σ^2 with confidence level at least 99%.

7. How would you generate uniformly distributed points on the surface of the 3-dimensional unit sphere?

8. Assume the Earth is perfectly spherical and you are standing somewhere on its surface. You travel exactly 1 mile south, then 1 mile east, then 1 mile north. Surprisingly, you find yourself back at the starting point. If you are not at the North Pole, where can you possibly be?!

9. Solve the Ornstein-Uhlenbeck SDE

$$dr_t = \lambda(\theta - r_t)dt + \sigma dW_t,$$

with $\lambda > 0$, which is used, e.g., in the Vasicek model for interest rates.

10. Find all the integer solutions of the equation

$$x^3 + y^3 = 2013.$$

11. Let X and Y be standard normal variables with joint normal distribution with correlation ρ. Find the expectation

$$\mathbb{E}\left[\operatorname{sgn}(X)\operatorname{sgn}(Y)\right],$$

where $\operatorname{sgn}(\cdot)$ is the sign function given by $\operatorname{sgn}(x) = 1$, if $x > 0$, $\operatorname{sgn}(x) = -1$, if $x < 0$, and $\operatorname{sgn}(0) = 0$.

12. How do you create a long Gamma, short vega options trading strategy?

13. Let X_t and Y_t be geometric Brownian motions driven by

$$\frac{dX_t}{X_t} = \mu_X dt + \sigma_X dW_t;$$
$$\frac{dY_t}{Y_t} = \mu_Y dt + \sigma_Y dB_t,$$

where W_t and B_t are correlated Brownian motions with constant correlation ρ. Show that

$$Z_t = \frac{X_t}{Y_t}$$

is also a geometric Brownian motion and determine its drift and volatility coefficients.

14. Find the k–th largest element in an unsorted array. Assume that k is always valid, i.e., $k \geq 1$ and k is less than or equal to the length of the array.

Note: You are looking for the k–th largest element in the sorted order, not the k–th distinct element of the array.

Example 1:

```
Input: [3,2,1,5,6,4] and k = 2
Output: 5
```

Example 2:

```
Input: [3,2,3,1,2,4,5,5,6] and k = 4
Output: 4
```

15. Given an array *nums*, there is a sliding window of size k which is moving from the very left of the array to the very right of the array. You can only see the k numbers in the window. Each time the sliding window moves right by one position. Assume that k is always valid, i.e., $k \geq 1$ and k is less than or equal to the size of the input array size for non-empty arrays.

    Write an algorithm that returns the maximum of the sliding window.

    Example:

```
Input: nums = [1,3,-1,-3,5,3,6,7], and k = 3
Output: [3,3,5,5,6,7]
```

Explanation:

```
Window Position                 Max
---------------------------    -----
[1  3  -1] -3  5  3  6  7        3
```

```
1 [3  -1  -3] 5  3  6  7        3
1  3 [-1  -3  5] 3  6  7        5
1  3  -1 [-3  5  3] 6  7        5
1  3  -1  -3 [5  3  6] 7        6
1  3  -1  -3  5 [3  6  7]       7
```

Solutions

Question 1. Nine months call options with strikes 20 and 25 on a non–dividend–paying underlying asset with spot price \$22 are trading for \$5.50 and \$1, respectively. Can you find an arbitrage?

Answer: Note that a call option with strike 0 on a non–dividend–paying underlying asset is the same as one unit of the asset, since the call with strike 0 will always be exercised at maturity by paying \$0, i.e., the strike of the option, to receive one unit of the asset. Thus, we are implicitly given a third call option with strike $K = 0$ and price \$22 (i.e., the spot price of the asset), and we can proceed to identify whether there is convexity arbitrage for these three call options.

Let $K_1 = 0$, $K_2 = 20$, $K_3 = 25$ and $C_1 = 22$, $C_2 = 5.50$, $C_3 = 1$. Note that $20 = \frac{1}{5} \cdot 0 + \frac{4}{5} \cdot 25$, i.e.,

$$K_2 = \frac{1}{5}K_1 + \frac{4}{5}K_3.$$

Since

$$\frac{1}{5}C_1 + \frac{4}{5}C_3 = 5.20 < 5.50 = C_2, \tag{1.1}$$

the convexity of option prices with respect to strike is violated.

The arbitrage strategy is to "buy low" $\frac{1}{5}C_1 + \frac{4}{5}C_3$ and "sell high" C_2. To normalize units, we multiply the positions by 500 to obtain the following arbitrage strategy: "buy low" $100C_1 + 400C_2$ and "sell high" $500C_2$. Note that buying $100C_1$, i.e., 100 calls with strike $K_1 = 0$, is equivalent to buying 100 units of the underlying asset since the asset does not pay dividends.

Arbitrage Strategy:
- buy 100 units of the underlying asset for \$2,200;
- buy 400 calls with strike $K_3 = 25$ for \$400;

- sell 500 calls with strike $K_2 = 20$ for \$2,750;
- realize a positive cash flow of \$150.

The positive cash flow \$150 represents risk–free profit since the arbitrage portfolio does not lose money at maturity:

The value of the arbitrage portfolio at the maturity T of the options is

$$\begin{aligned} V(T) &= 100S(T) - 500C_2(T) + 400C_3(T) \\ &= 100S(T) - 500\max(S(T) - 20, 0) \\ &\quad + 400\max(S(T) - 25, 0). \end{aligned}$$

If $S(T) \leq 20$,

$$V(T) = 100S(T) \geq 0.$$

If $20 < S(T) \leq 25$,

$$\begin{aligned} V(T) &= 100S(T) - 500(S(T) - 20) \\ &= 10000 - 400S(T) \\ &\geq 0. \end{aligned}$$

If $25 < S(T)$,

$$\begin{aligned} V(T) &= 100S(T) - 500(S(T) - 20) \\ &\quad + 400(S(T) - 25) \\ &= 0. \end{aligned}$$

Note that $150 = 500 \cdot (5.50 - 5.20)$, i.e., the risk–free profit \$150 is equal to the size of the convexity disparity \$5.50 – \$5.20 times the amplifier factor 500. $\square$

Question 2. (i) What is the sum of the eigenvalues of the correlation matrix of n random variables?

(ii) Find a lower bound for the sum of the eigenvalues of the inverse of a nonsingular correlation matrix of n random variables.

Answer: (i) The sum of the eigenvalues of a matrix is equal to the trace of the matrix, i.e., to the sum of the main diagonal entries of the matrix.[1] Since the correlation matrix of n random variables is an $n \times n$ matrix with all main diagonal entries equal to 1, the trace of the correlation matrix is equal to n. We conclude that the sum of the eigenvalues of the correlation matrix of n random variables is n.

(ii) If $\lambda_1, \lambda_2, \ldots, \lambda_n$ are the eigenvalues of the nonsingular $n \times n$ correlation matrix Ω, then $\lambda_i > 0$ for all $i = 1 : n$, since a nonsingular correlation matrix is symmetric positive definite. The eigenvalues of the inverse matrix Ω^{-1} are $\frac{1}{\lambda_1}, \frac{1}{\lambda_2}, \ldots, \frac{1}{\lambda_n}$. Thus, the question asks us what could be said about the sum

$$\sum_{i=1}^{n} \frac{1}{\lambda_i}$$

of the eigenvalues of Ω^{-1}.

Recall from (i) that the sum of the eigenvalues of the correlation matrix Ω is n, i.e.,

$$\sum_{i=1}^{n} \lambda_i = n. \tag{1.2}$$

Also, recall from the Cauchy–Schwartz inequality that

$$\left(\sum_{i=1}^{n} a_i^2\right) \left(\sum_{i=1}^{n} b_i^2\right) \geq \left(\sum_{i=1}^{n} a_i b_i\right)^2. \tag{1.3}$$

Since $\lambda_i > 0$ for all $i = 1 : n$, we can use the Cauchy–Schwartz inequality (1.3) for $a_i = \sqrt{\lambda_i}$ and $b_i = \frac{1}{\sqrt{\lambda_i}}$,

[1]This property follows from the fact that the eigenvalues of a matrix A are the roots of the characteristic polynomial $P_A(t) = \det(tI - A)$ of the matrix A; see, e.g., Theorem 4.1 from Stefanica [5].

$i = 1 : n$, to obtain that

$$\left(\sum_{i=1}^{n} \lambda_i\right) \left(\sum_{i=1}^{n} \frac{1}{\lambda_i}\right) \geq \left(\sum_{i=1}^{n} 1\right)^2 = n^2, \qquad (1.4)$$

since $a_i^2 = \left(\sqrt{\lambda_i}\right)^2 = \lambda_i$, $b_i^2 = \left(\frac{1}{\sqrt{\lambda_i}}\right)^2 = \frac{1}{\lambda_i}$, and $a_i b_i = \sqrt{\lambda_i} \cdot \frac{1}{\sqrt{\lambda_i}} = 1$, for all $i = 1 : n$.

From (1.4) and using (1.2), we find that

$$n \left(\sum_{i=1}^{n} \frac{1}{\lambda_i}\right) \geq n^2$$

and therefore conclude that

$$\sum_{i=1}^{n} \frac{1}{\lambda_i} \geq n.$$

In other words, the sum of the eigenvalues of the inverse of a nonsingular correlation matrix of n random variables is bounded from below by n. □

Question 3. Let W_t be a Wiener process, and let

$$X_t = \int_0^t W_\tau d\tau. \qquad (1.5)$$

What is the distribution of X_t? Is X_t a martingale?

Answer: Note that we can rewrite (1.5) in differential form as

$$dX_t = W_t dt = W_t dt + 0\, dW_t.$$

Then, X_t is a diffusion process with only drift part W_t, and therefore X_t is not a martingale.

We use integration by parts to find the distribution of X_t; a different solution can be found in Section 3.7.

By applying integration by parts, we obtain that

$$\begin{aligned} X_t &= \int_0^t W_\tau d\tau \\ &= tW_t - \int_0^t \tau dW_\tau \\ &= t\int_0^t dW_\tau - \int_0^t \tau dW_\tau \\ &= \int_0^t (t-\tau)dW_\tau. \end{aligned}$$

Recall that, if $f(t)$ is a deterministic square integrable function, then the stochastic integral $\int_0^t f(\tau)dW_\tau$ is a normal random variable of mean 0 and variance $\int_0^t |f(\tau)|^2 d\tau$, i.e.,

$$\int_0^t f(\tau)dW_\tau \sim N\left(0, \int_0^t |f(\tau)|^2 d\tau\right).$$

Thus,

$$\begin{aligned} X_t &= \int_0^t (t-\tau)dW_\tau \\ &\sim N\left(0, \int_0^t (t-\tau)^2\, d\tau\right) \\ &= N\left(0, \frac{t^3}{3}\right). \end{aligned}$$

We conclude that X_t is a normal random variable of mean 0 and variance $\frac{t^3}{3}$. □

Question 4. An 8×8 matrix contains zeros and ones. You may repeatedly choose any 3×3 or 4×4 block and flip all bits in the block (that is, convert zeros to ones, and ones to zeros). Can you always modify the original matrix into an all–zero matrix using these block flips?

Answer: No! Note that all the block flips are reversible, so it will suffice to show that there exist 8×8 matrices M

containing zeroes and ones that cannot be obtained using the block flips starting from an all–zero matrix.

Given a multiset of 3×3 and 4×4 blocks to be flipped in some order, the final matrix obtained is independent of the order in which the flips of the blocks in the multiset are applied. Moreover, we can remove all the block repetitions; in other words, we can reduce the multiset of the blocks flipped to a set with no repeated blocks by recognizing that flipping the same block twice does not affect the final matrix obtained at the end.

The total number of 3×3 blocks in an 8×8 matrix is 36: the upper left corner of the 3×3 block cannot be located in the 7–th or 8–th row or in the 7–th or 8–th column of the 8×8 matrix and therefore there are $6 \times 6 = 36$ possible positions for it. Similarly, the total number of 4×4 blocks in an 8×8 matrix is 25: the upper left corner of the 4×4 cannot be located in the 6–th, 7–th or 8–th row or in the 6–th, 7–th or 8–th column of the 8×8 matrix and therefore there are $5 \times 5 = 25$ possible positions for it.

Thus, there are $36+25 = 61$ blocks that can be flipped and the total number of different sets of blocks made with these 61 blocks (with no repeated blocks) is 2^{61}. Then, starting with an all–zero matrix, we can obtain at most 2^{61} distinct matrices. Since the total number of 8×8 matrices containing zeros and ones is 2^{64}, it follows that there exist matrices that cannot be obtained starting from an all–zero matrix by using block flips. $\square$

Question 5. Find all the values of ρ such that

$$\begin{pmatrix} 1 & 0.6 & -0.3 \\ 0.6 & 1 & \rho \\ -0.3 & \rho & 1 \end{pmatrix}$$

is a correlation matrix.

Answer: A symmetric matrix with diagonal entries equal to 1 is a correlation matrix if and only if the matrix is symmetric positive semidefinite. Thus, we need to find all the values of ρ such that the matrix

$$\Omega = \begin{pmatrix} 1 & 0.6 & -0.3 \\ 0.6 & 1 & \rho \\ -0.3 & \rho & 1 \end{pmatrix} \tag{1.6}$$

is symmetric positive semidefinite.

We give a short solution using Sylvester's criterion. Two more solutions, one using the Cholesky decomposition, and another one based on the definition of symmetric positive semidefinite matrices will be given in Section 3.2.

Recall from Sylvester's criterion that a matrix is symmetric positive semidefinite if and only if all its principal minors are greater than or equal to 0. Also, recall that the principal minors of a matrix are the determinants of all the square matrices obtained by eliminating the same rows and columns from the matrix. In particular, the matrix Ω from (1.6) has the following principal minors:

$$\det(1) = 1; \quad \det(1) = 1; \quad \det(1) = 1;$$

$$\begin{aligned} \det\begin{pmatrix} 1 & 0.6 \\ 0.6 & 1 \end{pmatrix} &= 0.64; \\ \det\begin{pmatrix} 1 & -0.3 \\ -0.3 & 1 \end{pmatrix} &= 0.91; \\ \det\begin{pmatrix} 1 & \rho \\ \rho & 1 \end{pmatrix} &= 1 - \rho^2; \end{aligned}$$

$$\begin{aligned} \det(\Omega) &= 1 - 0.36\rho - 0.09 - 0.36 - \rho^2 \\ &= 0.55 - 0.36\rho - \rho^2. \end{aligned}$$

Thus, it follows from Sylvester's criterion that Ω is a symmetric positive semidefinite matrix if and only if

$$\begin{aligned} 1-\rho^2 &\geq 0; \\ 0.55-0.36\rho-\rho^2 &\geq 0, \end{aligned}$$

which is equivalent to $-1 \leq \rho \leq 1$ and

$$\rho^2+0.36\rho-0.55 \leq 0. \tag{1.7}$$

Since the roots of the quadratic equation corresponding to (1.7) are -0.9432 and 0.5832, we conclude that the matrix Ω is symmetric positive semidefinite, and therefore a correlation matrix, if and only if

$$-0.9432 \leq \rho \leq 0.5832. \quad \square \tag{1.8}$$

Question 6. Given a sample of size 1 from the normal distribution with mean μ and variance σ^2, with both μ and σ are unknown, give a *finite* confidence interval for σ^2 with confidence level at least 99%.

Answer: Denote by X the single observation from the normal distribution with mean μ and variance σ^2, where both μ and σ are unknown. We construct a confidence interval $[0, T(X)]$ for σ^2, where $T(\cdot)$ denotes some statistic. This interval will be a confidence interval with confidence level at least 99% if for every μ and $\sigma^2 > 0$:

$$\mathbb{P}_{\mu,\sigma^2}\left(\sigma^2 > T(X)\right) < 0.01.$$

Note that the probability density function $f_X(x)$ of X satisfies

$$f_X(x) = \frac{1}{\sqrt{2\pi}\,\sigma}e^{-\frac{(x-\mu)^2}{2\sigma^2}} \leq \frac{1}{\sqrt{2\pi}\,\sigma}, \quad \forall\, x \in \mathbb{R}.$$

Then, for every $a \geq 0$, we have that

$$\begin{aligned} \mathbb{P}(|X| \leq a) &= \int_{-a}^{a} f_X(x)\,dx \leq \frac{2a}{\sqrt{2\pi}\,\sigma} \\ &\leq \frac{a}{\sigma}. \end{aligned} \tag{1.9}$$

By letting $a = 0.01\sigma$ in (1.9) we obtain that

$$\mathbb{P}(|X| \leq 0.01\sigma) \leq 0.01. \tag{1.10}$$

Note that

$$\begin{aligned} \mathbb{P}(|X| \leq 0.01\sigma) &= \mathbb{P}\left(X^2 \leq 0.0001\sigma^2\right) \\ &= \mathbb{P}\left(\sigma^2 \geq 10000X^2\right). \end{aligned} \tag{1.11}$$

From (1.10) and (1.11), we find that

$$\mathbb{P}\left(\sigma^2 \geq 10000X^2\right) \leq 0.01.$$

We conclude that $\left[0, 10000X^2\right]$ is a *finite* confidence interval for σ^2 with confidence level at least 99%. □

Question 7. How would you generate uniformly distributed points on the surface of the 3-dimensional unit sphere?

Answer: We will describe two different methods to accomplish this task.

Method 1: Spherical coordinates provide a mapping from every point $P(x, y, z)$ on the surface of the 3-dimensional unit sphere to a pair of angles (θ, ϕ), where $\theta \in [0, 2\pi]$ is the azimuthal angle and $\phi \in [0, \pi]$ is the polar angle, via transformations: $x = \sin(\phi)\cos(\theta)$, $y = \sin(\phi)\sin(\theta)$, $z = \cos(\phi)$.

A tempting way to try generating uniformly distributed points on the surface of the 3-dimensional unit sphere would be to generate both θ and ϕ angles uniformly at random from their respective intervals, and then apply the transformation above. However, this algorithm is incorrect, as the points generated by it will be clustered around the poles ($\phi = 0$ and $\phi = \pi$) while sparse around the equator ($\phi = \pi/2$).

Why is that so? The reason is that the Jacobian of the transformation above is equal to $\sin(\phi)$. In other words, the differential surface element dA in spherical coordinates is not $d\phi\, d\theta$, but rather $\sin(\phi) d\phi\, d\theta$. So, close to the poles of the sphere (i.e., when $\phi = 0$ or $\phi = \pi$), the differential surface element gets smaller as $\sin(\phi) \to 0$.

Our task is, hence, a bit more delicate: we have to find and then draw samples from a probability distribution with joint density $f(\theta, \phi)$ that maps from the (θ, ϕ)–plane to a uniform distribution on the unit sphere. Since for every $P(x, y, z)$ on the surface of the 3-dimensional unit sphere $f(P)$ has to be constant for a uniform distribution, we obtain that $f(P) = \frac{1}{4\pi}$, since the surface area of the unit sphere is 4π. Therefore,

$$f(P)\, dA = \frac{1}{4\pi}\, dA = f(\theta, \phi)\, d\theta\, d\phi.$$

Since $dA = \sin(\phi) d\phi\, d\theta$, it follows that

$$f(\theta, \phi) = \frac{1}{4\pi} \sin(\phi).$$

Integrating the joint density $f(\theta, \phi)$ to get the marginal densities of θ and ϕ separately, we find that

$$f(\theta) = \int_0^{\pi} f(\theta, \phi)\, d\phi = \frac{1}{2\pi},$$
$$f(\phi) = \int_0^{2\pi} f(\theta, \phi)\, d\theta = \frac{\sin(\phi)}{2}.$$

Clearly, θ is uniformly distributed over $[0, 2\pi]$, and, hence, θ can be sampled as 2π times the output from a readily available uniform random generator in $[0, 1]$. How do we, however, use the same generator to sample ϕ from a probability distribution with density $f(\phi) = \frac{\sin(\phi)}{2}$? We use the inverse transform sampling method.

Note that the cumulative distribution function (cdf) for the distribution of ϕ is

$$F(\phi) = \int_0^{\phi} f(s)\, ds = \frac{1}{2}\left(1 - \cos(\phi)\right).$$

The function $F(\phi)$ is strictly increasing from $[0, \pi]$ to $[0, 1]$, and, as such, has an inverse function $F^{-1}(u)$ given by

$$F^{-1}(u) = \arccos(1 - 2u).$$

Let $U \sim U[0, 1]$ be uniformly distributed over $[0, 1]$. Then,

$$\mathbb{P}(U \leq F(\phi)) = F(\phi),$$

and

$$\mathbb{P}\left(F^{-1}(U) \leq \phi\right) = F(\phi).$$

Therefore, $F(\phi)$ is the cdf of the random variable $F^{-1}(U)$. In other words, $F^{-1}(U)$ has the same probability distribution as ϕ. Hence, to sample ϕ from a probability distribution with density $f(\phi) = \frac{\sin(\phi)}{2}$, we generate a uniform random number U from $[0, 1]$ using a readily available uniform random generator, then compute $\phi = \arccos(1 - 2U)$.

Finally, once we have sampled θ and ϕ, then x, y, and z are computed using the spherical transformation.

Method 2: Assume that we have available a random generator from the standard normal distribution, such as the Box–Muller method; see, e.g., Glasserman [2]. Generate three dependent standard normally distributed numbers X, Y, and Z to form a vector $\vec{v} = (X, Y, Z)$. Intuitively, this vector will point in a uniformly random direction in the 3-dimensional space. Next, we normalize the vector by dividing it by its norm, to obtain the point $P = \left(\frac{X}{\|\vec{v}\|}, \frac{Y}{\|\vec{v}\|}, \frac{Z}{\|\vec{v}\|}\right)$ on the unit sphere. In order to show that P is uniformly distributed on the surface of the unit sphere, it suffices to prove that $\vec{v}$ truly points in a uniformly random direction.

As X, Y, and Z are each sampled independently from the standard normal distribution, the probability density function of (X, Y, Z) is given by the product of their marginal densities:

$$\begin{aligned} f(x,y,z) &= f(\vec{v}) \\ &= \left(\frac{1}{\sqrt{2\pi}}e^{-\frac{1}{2}x^2}\right)\left(\frac{1}{\sqrt{2\pi}}e^{-\frac{1}{2}y^2}\right)\left(\frac{1}{\sqrt{2\pi}}e^{-\frac{1}{2}z^2}\right) \\ &= \frac{1}{(2\pi)^{3/2}}e^{-\frac{1}{2}(x^2+y^2+z^2)} \\ &= \frac{1}{(2\pi)^{3/2}}e^{-\frac{1}{2}\|\vec{v}\|^2}. \end{aligned}$$

In other words, the probability density function of $\vec{v}$ depends only on its norm and not on any angles such as θ or ϕ. In conclusion, by finding where the ray $\vec{v}$ intersects the unit sphere, we obtain a sample from a uniform distribution on the surface of the unit sphere. □

Question 8. Assume the Earth is perfectly spherical and you are standing somewhere on its surface. You travel exactly 1 mile south, then 1 mile east, then 1 mile north. Surprisingly, you find yourself back at the starting point. If you are not at the North Pole, where can you possibly be?!

Answer: There are infinitely many locations, aside from the North Pole, that have this property.

Somewhere near the South Pole, there is a latitude that has a circumference of one mile. In other words, if you are at this latitude and start walking east (or west), in one mile you will be back exactly where you started from. If you instead start at some point one mile north of this latitude, your journey will take you one mile south to this special latitude, then one mile east "around the globe" and finally one mile north right back to wherever you started from. Moreover, there are infinitely many points on the Earth that are one mile north of this special latitude, where you could start your journey and eventually end up exactly where you started.

We are still not finished! There are infinitely many special latitudes as well; namely, you could start at any point one mile north of the latitude that has a circumference of $1/k$ miles, where k is a positive integer. Your journey will take you one mile south to this special latitude, then one mile east looping "around the globe" k times, and finally one mile north right back to where you started from. □

Question 9. Solve the Ornstein-Uhlenbeck SDE

$$dr_t = \lambda(\theta - r_t)dt + \sigma dW_t, \tag{1.12}$$

with $\lambda > 0$, which is used, e.g., in the Vasicek model for interest rates.

Answer: We can rewrite (1.12) as

$$dr_t + \lambda r_t dt = \lambda\theta dt + \sigma dW_t. \tag{1.13}$$

By multiplying (1.13) on both sides by the integrating factor $e^{\lambda t}$, we obtain that

$$e^{\lambda t} dr_t + \lambda e^{\lambda t} r_t dt = \lambda\theta e^{\lambda t} dt + \sigma e^{\lambda t} dW_t,$$

which is equivalent to

$$d\left(e^{\lambda t} r_t\right) = \lambda\theta e^{\lambda t} dt + \sigma e^{\lambda t} dW_t. \tag{1.14}$$

By integrating (1.14) from 0 to t, it follows that

$$\begin{aligned} e^{\lambda t} r_t - r_0 &= \lambda\theta \int_0^t e^{\lambda s} ds + \sigma \int_0^t e^{\lambda s} dW_s \\ &= \theta\left(e^{\lambda t} - 1\right) + \sigma \int_0^t e^{\lambda s} dW_s. \end{aligned}$$

By solving for r_t, we find that the solution to the Ornstein-Uhlenbeck SDE is

$$\begin{aligned} r_t &= e^{-\lambda t} r_0 + e^{-\lambda t}\theta\left(e^{\lambda t} - 1\right) + \sigma e^{-\lambda t} \int_0^t e^{\lambda s} dW_s \\ &= e^{-\lambda t} r_0 + \theta\left(1 - e^{-\lambda t}\right) + \sigma \int_0^t e^{-\lambda(t-s)} dW_s. \end{aligned}$$

Note that the process r_t is mean reverting to θ, regardless of the starting point r_0. To see this, recall that the expected value of the stochastic integral $\int_0^t f(s)dW_s$ of a non-random function $f(s)$ is 0. Then,

$$E\left[\int_0^t e^{-\lambda(t-s)}dW_s\right] = 0,$$

and therefore

$$E[r_t] \;=\; e^{-\lambda t}r_0 + \theta\left(1 - e^{-\lambda t}\right).$$

Thus,

$$\lim_{t\to\infty} E[r_t] \;=\; \theta. \quad \square$$

Question 10. Find all the integer solutions of the equation

$$x^3 + y^3 \;=\; 2013.$$

Answer: The equation has no integer solutions.

The challenge in this problem comes from the fact that we can write the equation as

$$(x+y)(x^2 - xy + y^2) \;=\; 3 \cdot 11 \cdot 61,$$

which means that $x + y$ has 16 possible values which are the positive and negative divisors of $2013 = 3 \cdot 11 \cdot 61$. This would lead to a long–winded solution.

However, there is a straightforward way to see that the equation has no integer solutions, by looking at residuals modulo 9.

Note that $2013 \equiv 6 \pmod 9$; here,

$$a \equiv b \pmod m \iff m \mid (a-b),$$

where $m > 1$ is a positive integer and a and b are integers.

Furthermore:

- if $a \equiv 0 \pmod 3$, then $a^3 \equiv 0 \pmod 9$;

- if $a \equiv 1 \pmod 3$, then $a^3 \equiv 1 \pmod 9$;
- if $a \equiv 2 \pmod 3$, then $a^3 \equiv 8 \pmod 9$.

This means that $x^3 + y^3$ can only be equal to 0, 1, 2, 7 or 8 modulo 9, and cannot be equal to 2013 for any integers x and y since $2013 \equiv 6 \pmod 9$. □

Question 11. Let X and Y be standard normal variables with joint normal distribution with correlation ρ. Find the expectation

$$\mathbb{E}\left[\operatorname{sgn}(X)\operatorname{sgn}(Y)\right],$$

where $\operatorname{sgn}(\cdot)$ is the sign function given by $\operatorname{sgn}(x) = 1$, if $x > 0$, $\operatorname{sgn}(x) = -1$, if $x < 0$, and $\operatorname{sgn}(0) = 0$.

Answer: If $\rho = 1$, then

$$\mathbb{E}\left[\operatorname{sgn}(X)\operatorname{sgn}(Y)\right] = \mathbb{E}\left[\operatorname{sgn}(Z)^2\right] = \mathbb{E}\left[1\right] = 1, \qquad (1.15)$$

where Z is the standard normal variable, and, if $\rho = -1$,

$$\mathbb{E}\left[\operatorname{sgn}(X)\operatorname{sgn}(Y)\right] = \mathbb{E}\left[-\operatorname{sgn}(Z)^2\right] = \mathbb{E}\left[-1\right] = -1. \qquad (1.16)$$

If $\rho \in (-1, 1)$, we obtain that

$$\begin{aligned} & \mathbb{E}\left[\operatorname{sgn}(X)\operatorname{sgn}(Y)\right] \\ = \ & \mathbb{P}\left[X > 0, Y > 0\right] + \mathbb{P}\left[X < 0, Y < 0\right] \\ & - \mathbb{P}\left[X > 0, Y < 0\right] - \mathbb{P}\left[X < 0, Y > 0\right]. \end{aligned} \qquad (1.17)$$

Note that

$$\begin{aligned} \mathbb{P}\left[X > 0, Y > 0\right] &= \mathbb{P}\left[X < 0, Y < 0\right]; \qquad (1.18) \\ \mathbb{P}\left[X > 0, Y < 0\right] &= \mathbb{P}\left[X < 0, Y > 0\right], \qquad (1.19) \end{aligned}$$

due to symmetry, and therefore (1.17) can be written using (1.18–1.19) as

$$\begin{aligned} & \mathbb{E}\left[\operatorname{sgn}(X)\operatorname{sgn}(Y)\right] \\ = \ & 2\,\mathbb{P}\left[X > 0, Y > 0\right] - 2\,\mathbb{P}\left[X > 0, Y < 0\right] \end{aligned} \qquad (1.20)$$

Moreover,

$$\begin{aligned} & \mathbb{P}[X>0,Y>0]+\mathbb{P}[X<0,Y<0] \\ & +\mathbb{P}[X>0,Y<0]+\mathbb{P}[X<0,Y>0] \\ = \;& 1. \end{aligned} \tag{1.21}$$

Using (1.18–1.19) in (1.21), we find that

$$2\,\mathbb{P}[X>0,Y>0]+2\,\mathbb{P}[X>0,Y<0] \;=\; 1$$

and therefore

$$\mathbb{P}[X>0,Y<0] \;=\; \frac{1}{2}-\mathbb{P}[X>0,Y>0]. \tag{1.22}$$

By substituting (1.22) in (1.20), we obtain that

$$\mathbb{E}[\operatorname{sgn}(X)\operatorname{sgn}(Y)] \;=\; 4\,\mathbb{P}[X>0,Y>0]-1. \tag{1.23}$$

To compute $\mathbb{P}[X>0,Y>0]$, recall that, if X and Y are standard normal variables with joint normal distribution with correlation ρ, then there exist two independent standard normal variables Z_1 and Z_2 such that

$$\begin{pmatrix} X \\ Y \end{pmatrix} \;=\; \begin{pmatrix} Z_1 \\ \rho Z_1+\sqrt{1-\rho^2}\,Z_2 \end{pmatrix}. \tag{1.24}$$

Let $\widetilde{\rho}=\sqrt{1-\rho^2}$. From (1.24), we obtain that

$$Y \;=\; \rho X+\widetilde{\rho}Z, \tag{1.25}$$

where we denoted Z_2 by Z for simplicity. Note that $X = Z_1$ and $Z = Z_2$ are independent standard normals.

Then, from (1.25) and using the fact that X and Z are independent standard normal variables, it follows that

$$\begin{aligned} & \mathbb{P}[X>0,Y>0] \\ = \;& \mathbb{P}[X>0,\rho X+\widetilde{\rho}Z>0] \\ = \;& \mathbb{P}\left[X>0,Z>-\frac{\rho}{\widetilde{\rho}}X\right] \\ = \;& \frac{1}{2\pi}\int_0^\infty\int_{-\frac{\rho}{\widetilde{\rho}}x}^\infty e^{-\frac{x^2+z^2}{2}}\,dzdx. \end{aligned} \tag{1.26}$$

We use a polar coordinates change of variables to compute the integral (1.26). Let

$$x = r\cos(\theta); \quad z = r\sin(\theta),$$

and recall that

$$dzdx = rd\theta dr. \tag{1.27}$$

Note that

$$\begin{aligned} & -\frac{\rho}{\widetilde{\rho}}x < z < \infty \\ \iff & -\frac{\rho}{\widetilde{\rho}} < \tan(\theta) < \infty \\ \iff & \alpha < \theta < \frac{\pi}{2}, \end{aligned} \tag{1.28}$$

where

$$\alpha = \arctan\left(-\frac{\rho}{\widetilde{\rho}}\right).$$

Note that α is the signed angle between the x-axis and the straight line $\rho x + \widetilde{\rho} z = 0$ on the (x, z) plane.

From (1.26) and using (1.27) and (1.28), we obtain that

$$\begin{aligned} & \mathbb{P}[X > 0, Y > 0] \\ = & \frac{1}{2\pi}\int_0^\infty \int_{-\frac{\rho}{\widetilde{\rho}}x}^\infty e^{-\frac{x^2+z^2}{2}}\, dzdx \\ = & \frac{1}{2\pi}\int_0^\infty \int_\alpha^{\frac{\pi}{2}} e^{-\frac{r^2}{2}} r\, d\theta dr \\ = & \frac{1}{2\pi}\left(\frac{\pi}{2} - \alpha\right)\int_0^\infty re^{-\frac{r^2}{2}}\, dr \\ = & \left(\frac{1}{4} - \frac{\alpha}{2\pi}\right)\left(-e^{-\frac{r^2}{2}}\right)\Big|_0^\infty \\ = & \frac{1}{4} - \frac{\alpha}{2\pi}. \end{aligned} \tag{1.29}$$

From (1.23) and (1.29), we conclude that

$$\begin{aligned}\mathbb{E}\left[\operatorname{sgn}(X)\operatorname{sgn}(Y)\right] &= 4\,\mathbb{P}\left[X>0, Y>0\right]-1 \\ &= 4\left(\frac{1}{4}-\frac{\alpha}{2\pi}\right)-1 \\ &= -\frac{2\alpha}{\pi}. \end{aligned} \tag{1.30}$$

Formulas (1.15) and (1.16) for $\mathbb{E}\left[\operatorname{sgn}(X)\operatorname{sgn}(Y)\right]$ corresponding to $\rho = 1$ and $\rho = -1$, respectively, can be obtained from the general formula (1.30) as limiting cases when ρ goes to 1 and to -1. For example,

$$\begin{aligned} &\lim_{\rho\searrow -1} \mathbb{E}\left[\operatorname{sgn}(X)\operatorname{sgn}(Y)\right] \\ &= \lim_{\rho\searrow -1}\left(-\frac{2\alpha}{\pi}\right) \\ &= -\frac{2}{\pi}\lim_{\rho\searrow -1}\arctan\left(-\frac{\rho}{\widetilde{\rho}}\right) \\ &= -\frac{2}{\pi}\lim_{\rho\searrow -1}\arctan\left(-\frac{\rho}{\sqrt{1-\rho^2}}\right) \\ &= -\frac{2}{\pi}\cdot\frac{\pi}{2} \\ &= -1, \end{aligned}$$

which is the same as (1.16), since

$$\lim_{\rho\searrow -1}\left(-\frac{\rho}{\sqrt{1-\rho^2}}\right) = \infty$$

and therefore

$$\lim_{\rho\searrow -1}\arctan\left(-\frac{\rho}{\sqrt{1-\rho^2}}\right) = \frac{\pi}{2}. \quad \square$$

Question 12. How do you create a long Gamma, short vega options trading strategy?

Answer: Both Gamma and vega are highest for options around at–the–money (ATM). However, the Gamma of ATM options is higher for options with shorter maturity (i.e., for short–dated options), while the vega of ATM options is higher for longer maturity options (i.e., for long–dated options); see Figure 1.1 and Figure 1.2, respectively.

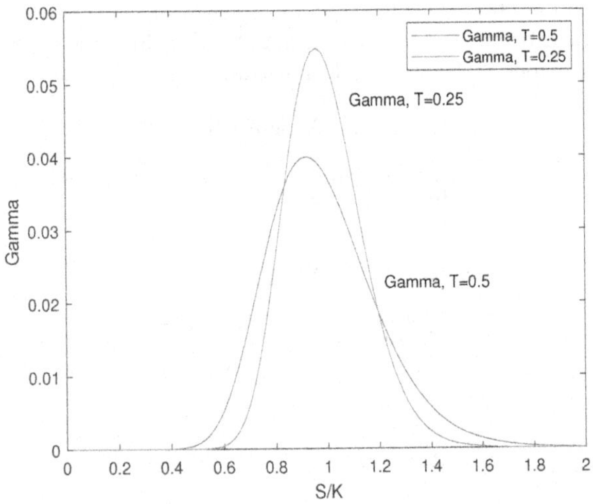

Figure 1.1: Dependence of Gamma on time to maturity

A trader who buys short–dated ATM options and sells the same number of long–dated ATM options will be long Gamma and short vega.

Note that calls and puts with the same strike have

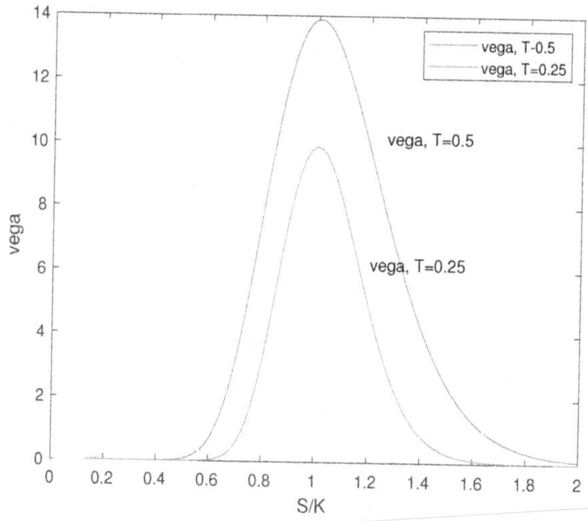

Figure 1.2: Dependence of vega on time to maturity

the same Gamma and vega, a consequence of the Put–Call parity, so you can take positions in either call or put options.

Also, the long Gamma, short vega portfolio can be made Delta–neutral by taking an appropriate position in the underlying asset. The delta of short–dated ATM options is smaller than the delta of long–dated ATM options. Then, the delta of the long Gamma, short vega portfolio is negative and therefore the trader will have to purchase units of the underlying asset in order to make the portfolio Delta–neutral. □

Question 13. Let X_t and Y_t be geometric Brownian

motions driven by

$$\frac{dX_t}{X_t} = \mu_X dt + \sigma_X dW_t; \tag{1.31}$$

$$\frac{dY_t}{Y_t} = \mu_Y dt + \sigma_Y dB_t, \tag{1.32}$$

where W_t and B_t are correlated Brownian motions with constant correlation ρ. Show that

$$Z_t = \frac{X_t}{Y_t}$$

is also a geometric Brownian motion and determine its drift and volatility coefficients.

Answer: Let

$$f(x,y) = \frac{x}{y}.$$

By applying Itô's lemma to $Z_t = \frac{X_t}{Y_t}$, we obtain that

$$\begin{aligned}
dZ_t &= d\left(\frac{X_t}{Y_t}\right) = d\,f(X_t,Y_t) \\
&= \frac{\partial f}{\partial x}(X_t,Y_t)\,dX_t + \frac{\partial f}{\partial y}(X_t,Y_t)\,dY_t \\
&\quad + \frac{1}{2}\frac{\partial^2 f}{\partial x^2}(X_t,Y_t)\,d[X]_t + \frac{1}{2}\frac{\partial^2 f}{\partial y^2}(X_t,Y_t)\,d[Y]_t \\
&\quad + \frac{\partial^2 f}{\partial x \partial y}(X_t,Y_t)\,d[X,Y]_t.
\end{aligned}$$

Note that

$$\frac{\partial f}{\partial x}(x,y) = \frac{1}{y}; \quad \frac{\partial f}{\partial y}(x,y) = -\frac{x}{y^2};$$

$$\frac{\partial^2 f}{\partial x^2}(x,y) = 0; \quad \frac{\partial^2 f}{\partial y^2}(x,y) = \frac{2x}{y^3}; \quad \frac{\partial^2 f}{\partial x \partial y}(x,y) = -\frac{1}{y^2}$$

and therefore

$$dZ_t = \frac{1}{Y_t}\,dX_t - \frac{X_t}{Y_t^2}\,dY_t + \frac{X_t}{Y_t^3}\,d[Y]_t - \frac{1}{Y_t^2}\,d[X,Y]_t. \tag{1.33}$$

Since

$$\begin{aligned} d[Y]_t &= \sigma_Y^2 Y_t^2 dt; \\ d[X,Y]_t &= \rho\sigma_X\sigma_Y X_t Y_t dt, \end{aligned}$$

it follows from (1.33) that

$$\begin{aligned} dZ_t &= \frac{1}{Y_t}\,dX_t - \frac{X_t}{Y_t^2}\,dY_t + \frac{X_t}{Y_t}\sigma_Y^2\,dt - \frac{X_t}{Y_t}\rho\sigma_X\sigma_Y\,dt \\ &= \frac{X_t}{Y_t}\frac{dX_t}{X_t} - \frac{X_t}{Y_t}\frac{dY_t}{Y_t} + \frac{X_t}{Y_t}\sigma_Y^2\,dt - \frac{X_t}{Y_t}\rho\sigma_X\sigma_Y\,dt \\ &= Z_t\,\frac{dX_t}{X_t} - Z_t\,\frac{dY_t}{Y_t} + Z_t\sigma_Y^2\,dt - Z_t\rho\sigma_X\sigma_Y\,dt, \end{aligned}$$

which can be written as

$$\frac{dZ_t}{Z_t} = \frac{dX_t}{X_t} - \frac{dY_t}{Y_t} + \left(\sigma_Y^2 - \rho\sigma_X\sigma_Y\right)\,dt. \tag{1.34}$$

By substituting (1.31) and (1.32) in (1.34), we obtain that

$$\begin{aligned} \frac{dZ_t}{Z_t} &= (\mu_X dt + \sigma_X dW_t) - (\mu_Y dt + \sigma_Y dB_t) \\ &\quad + \left(\sigma_Y^2 - \rho\sigma_X\sigma_Y\right)\,dt \\ &= \left(\mu_X - \mu_Y + \sigma_Y^2 - \rho\sigma_X\sigma_Y\right)\,dt \\ &\quad + (\sigma_X dW_t - \sigma_Y dB_t) \\ &= \mu_Z\,dt\; +\; (\sigma_X dW_t - \sigma_Y dB_t)\,, \end{aligned} \tag{1.35}$$

where

$$\mu_Z = \mu_X - \mu_Y + \sigma_Y^2 - \rho\sigma_X\sigma_Y.$$

Note that $\widetilde{W}_t$ given by

$$d\widetilde{W}_t = \frac{\sigma_X dW_t - \sigma_Y dB_t}{\sqrt{\sigma_X^2 - 2\rho\sigma_X\sigma_Y + \sigma_Y^2}}$$

is a Brownian motion, and let

$$\sigma_Z = \sqrt{\sigma_X^2 - 2\rho\sigma_X\sigma_Y + \sigma_Y^2}.$$

Then,

$$\sigma_X dW_t - \sigma_Y dB_t = \sigma_Z d\widetilde{W}_t, \qquad (1.36)$$

and we conclude from (1.35) and (1.36) that Z_t satisfies the SDE

$$\frac{dZ_t}{Z_t} = \mu_Z\, dt + \sigma_Z\, d\widetilde{W}_t.$$

Thus, Z_t is a geometric Brownian motion with drift μ_Z and volatility σ_Z, where

$$\mu_Z = \mu_X - \mu_Y + \sigma_Y^2 - \rho\sigma_X\sigma_Y;$$
$$\sigma_Z = \sqrt{\sigma_X^2 - 2\rho\sigma_X\sigma_Y + \sigma_Y^2}. \quad \square$$

Question 14. Find the k–th largest element in an unsorted array. Assume that k is always valid, i.e., $k \geq 1$ and k is less than or equal to the length of the array.

Note: You are looking for the k–th largest element in the sorted order, not the k–th distinct element of the array.

Example 1:

```
Input: [3,2,1,5,6,4] and k = 2
Output: 5
```

Example 2:

```
Input: [3,2,3,1,2,4,5,5,6] and k = 4
Output: 4
```

Answer:
Solution 1: Use a max heap data structure as follows (sample code in C++):

```
class Solution {
public:
  int findKthLargest(vector<int>& nums, int k) {
    std::priority_queue<int> max_heap;
    for (int i = 0; i < nums.size(); ++i){
      max_heap.push(nums[i]);
```

```
    }
    int j = 0;
    while (j++ < k - 1){
      max_heap.pop();
    }
    return max_heap.top();
  }
};
```

Solution 2: Use a quick selection algorithm as follows (sample code in C++):

```
class Solution {
public:
  int findKthLargest(vector<int>& nums, int k) {
    const int size_n = nums.size();
    int left = 0, right = size_n;
    whilo (lefL < right) {
      int i = left, j = right - 1, pivot = nums[left];
      while(i <= j) {
        while (i <= j && nums[i] >= pivot) i++;
        while (i <= j && nums[j] < pivot) j--;
        if (i < j)
          swap(nums[i++], nums[j--]);
      }
      swap(nums[left], nums[j]);
      if (j == k - 1) return nums[j];
      if (j < k - 1) left = j + 1;
      else right = j;
    }
  }
};
```

Question 15. Given an array *nums*, there is a sliding window of size k which is moving from the very left of the array to the very right of the array. You can only see the k numbers in the window. Each time the sliding window moves right by one position. Assume that k is always valid, i.e., $k \geq 1$ and k is less than or equal to the size of the input array size for non-empty arrays.

Write an algorithm that returns the maximum of the sliding window.

Example:

```
Input: nums = [1,3,-1,-3,5,3,6,7], and k = 3
Output: [3,3,5,5,6,7]
```

Explanation:

```
Window Position                Max
---------------------------   -----
[1  3  -1] -3  5  3  6  7       3
 1 [3  -1  -3] 5  3  6  7       3
 1  3 [-1  -3  5] 3  6  7       5
 1  3  -1 [-3  5  3] 6  7       5
 1  3  -1  -3 [5  3  6] 7       6
 1  3  -1  -3  5 [3  6  7]      7
```

Answer: Use a deque (double-ended queue) data structure as follows (sample code in C++):

```
class Solution {
public:
vector<int> maxSlidingWindow(vector<int>& nums, int k) {
 int n = nums.size();
 vector<int> res;
 if (n == 0) return res;
 if (k == 1) return nums;
 deque<int> myDeque;
 for (int i = 0; i < n; ++i){
  if (myDeque.empty()) myDeque.push_back(i);
  else {
   if (i - myDeque.front() == k) myDeque.pop_front();
   if (i - myDeque.front() < k){
    if (nums[myDeque.back()] > nums[i]) myDeque.push_back(i);
    if (nums[myDeque.back()] < nums[i]){
     while (!myDeque.empty() && nums[i] > nums[myDeque.back()]){
      myDeque.pop_back();
     }
     myDeque.push_back(i);
    }
   }
  }
```

```
  if (i >= k-1) res.push_back(nums[myDeque.front()]);
 }
 return res;
}
};
```

Chapter 2

Questions

1. Mathematics, calculus, differential equations.

2. Covariance and correlation matrices. Linear algebra.

3. Financial instruments: options, bonds, swaps, forwards, futures.

4. C++. Data structures.

5. Monte Carlo simulations. Numerical methods.

6. Probability. Stochastic calculus.

7. Brainteasers.

2.1 Mathematics, calculus, differential equations.

1. What is the value of i^i, where $i = \sqrt{-1}$?

2. Which number is larger, π^e or e^π?

3. Show that
$$\frac{e^x + e^y}{2} \geq e^{\frac{x+y}{2}}, \quad \forall\, x, y \in \mathbb{R}.$$

4. Solve $x^6 = 64$.

5. What is the derivative of x^x?

6. Calculate
$$\sqrt{2 + \sqrt{2 + \sqrt{2 + \ldots}}}.$$

7. Find x such that
$$x^{x^{x^{\cdot^{\cdot^{\cdot}}}}} = 2.$$

8. Which of the following series converge:
$$\sum_{k=1}^{\infty} \frac{1}{k}; \quad \sum_{k=1}^{\infty} \frac{1}{k^2}; \quad \sum_{k=2}^{\infty} \frac{1}{k \ln(k)}?$$

9. Compute

$$\int \frac{1}{1+x^2}\, dx.$$

10. Compute

$$\int x\ln(x)dx \quad \text{and} \quad \int xe^x\, dx.$$

11. Compute

$$\int x^n \ln(x)\ dx.$$

12. Compute

$$\int (\ln(x))^n\ dx.$$

13. Solve the ODE

$$y'' - 4y' + 4y = 1.$$

14. Find $f(x)$ such that

$$f'(x) \;=\; f(x)(1 - f(x)).$$

15. Derive the Black-Scholes PDE.

16. Calculate the limit

$$\lim_{x\to 0} \frac{(\cos(2x) - 1)^2 (\ln(1+3x) - 3x)}{x^3(e^{x^3} - 1)}$$

17. (i) Prove that

$$\sum_{i=1}^{n} i^3 = \left(\frac{n(n+1)}{2}\right)^2,$$

for all positive integers n.

(ii) Derive the formula

$$\sum_{i=1}^{n} i^3 = \left(\frac{n(n+1)}{2}\right)^2.$$

18. Prove the Cauchy–Schwartz inequality

$$\left(\sum_{i=1}^{n} x_i^2\right)\left(\sum_{i=1}^{n} y_i^2\right) \geq \left(\sum_{i=1}^{n} x_i y_i^2\right)^2,$$

where x_i and y_i are real numbers, for $i = 1 : n$.

19. Compute

$$\int_{-\infty}^{\infty} e^{-x^2}\, dx.$$

20. Solve the differential equation

$$y'(x) - y(x) = f(x),$$

where $f(x)$ is a continuous function.

21. Solve the differential equation

$$y''(x) - y(x) = g(x),$$

where $g(x)$ is a continuous function.

22. Find the value of

$$\int_0^a \frac{f(x)}{f(x) + f(a-x)}\, dx,$$

assuming that $f(x)$ is a function such that the integral above exists.

2.2 Covariance and correlation matrices. Linear algebra.

1. Show that any covariance matrix is symmetric positive semidefinite. Show that the same is true for correlation matrices.

2. Find the correlation matrix of three random variables with covariance matrix

$$\Sigma_X = \begin{pmatrix} 1 & 0.36 & -1.44 \\ 0.36 & 4 & 0.80 \\ -1.44 & 0.80 & 9 \end{pmatrix}.$$

3. Assume that all the entries of an $n \times n$ correlation matrix which are not on the main diagonal are equal to ρ. Find upper and lower bounds on the possible values of ρ.

4. How many eigenvalues does an $n \times n$ matrix with real entries have? How many eigenvectors?

5. Let

$$A = \begin{pmatrix} 2 & -2 \\ -2 & 5 \end{pmatrix}.$$

(i) Find a 2×2 matrix M such that $M^2 = A$;

(ii) Find a 2×2 matrix M such that $A = MM^t$.

6. The 2×2 matrix A has eigenvalues 2 and -3 with corresponding eigenvectors $\begin{pmatrix} 1 \\ 2 \end{pmatrix}$ and $\begin{pmatrix} -1 \\ 3 \end{pmatrix}$. If $v = \begin{pmatrix} 3 \\ 1 \end{pmatrix}$, find Av.

7. Let A and B be square matrices of the same size. Show that the traces of the matrices AB and BA are equal.

8. Can you find $n \times n$ matrices A and B such that

$$AB - BA = I_n,$$

where I_n is the identity matrix of size n?

9. A probability matrix is a matrix with nonnegative entries such that the sum of the entries in each row of the matrix is equal to 1. Show that the product of two probability matrices is a probability matrix.

10. Find all the values of ρ such that

$$\begin{pmatrix} 1 & 0.6 & -0.3 \\ 0.6 & 1 & \rho \\ -0.3 & \rho & 1 \end{pmatrix}$$

is a correlation matrix.[1]

11. Let

$$A = \begin{pmatrix} 2 & -1 \\ -1 & 2 \end{pmatrix}.$$

[1] A solution to this question was given in Chapter 1 using Sylvester's criterion; two different solutions will be given herein.

Compute A^{50}.

12. Let A be an $n \times n$ matrix. How do you compute A^{50} in a numerically efficient way minimizing the roundoff errors? What is the operation count to do so?

13. Let X_1, X_2, and X_3 be three random variables on the same probability space. If the correlation between X_1 and X_3 is 0, what is the possible range for the correlations between X_1 and X_2 and between X_2 and X_3?

14. Let A and B be square matrices of the same size such that $AB = BA$ and $B^3 = 0$. If the matrix $I + AB$ is nonsingular, find the inverse matrix of $I + AB$.

2.3 Financial instruments: options, bonds, swaps, forwards, futures.

1. The prices of three put options with strikes 40, 50, and 70, but otherwise identical, are \$10, \$20, and \$30, respectively. Is there an arbitrage opportunity present? If yes, how can you make a riskless profit?

2. The price of a stock is \$50. In three months, it will either be \$47 or \$52, with 50% probability. How much would you pay for an at–the–money put? Assume for simplicity that the stock pays no dividends and that interest rates are zero.

3. A stock worth \$50 today will be worth either \$60 or \$40 in three months, with equal probability. The value of a three months at–the–money put on this stock is \$4. Does the value of the three months ATM put increase or decrease, and by how much, if the probability of the stock going up to \$60 were 75% and the probability of the stock going down to \$40 were 25%?

4. What is risk–neutral pricing?

5. Describe briefly how you arrive at the Black–Scholes formula.

6. How much should a three months at–the–money put on an asset with spot price \$40 and volatility 30% be worth? Assume, for simplicity, that interest rates are zero and that the asset does not pay dividends.

7. If the price of a stock doubles in one day, by how much will the value of a call option on this stock change?

8. What are the smallest and largest values that Delta can take?

9. What is the Delta of an at–the–money call? What is the Delta of an at–the–money put?

10. What is the Put–Call parity? How do you prove it?

11. Show that the time value of a European call option is highest at–the–money.

12. What is implied volatility? What is a volatility smile? How about a volatility skew?

13. What is the Gamma of an option? Why is it preferable to have small Gamma? Why is the Gamma of plain vanilla options positive?

14. When are a European call and a European put worth the same? (The options are written on the same asset and have the same strike and maturity.) What is the intuition behind this result?

15. What is the two year volatility of an asset with 30% six months volatility?

16. How do you value an interest rate swap?

17. By how much will the price of a ten year zero coupon bond change if the yield increases by ten basis points?

18. A five year bond with 3.5 years duration is worth 102. What is the value of the bond if the yield decreases by fifty basis points?

19. What is a forward contract? What is the forward price?

20. What is the forward price for treasury futures contracts? What is the forward price for commodities futures contracts?

21. What is a Eurodollar futures contract?

22. What are the most important differences between forward contracts and futures contracts?

23. What is the ten–day 99% VaR of a portfolio with a five–day 98% VaR of $10 million?

24. Put options with strikes 30 and 20 on the same underlying asset and with the same maturity are trading for $6 and $4, respectively. Can you find an arbitrage?

25. I sell a one month put option with 28% implied volatility today and I hedge my position "continuously" until maturity. In one month, I calculate that the realized volatility of the underlying asset was 16%. Did I make money or did I lose money?

26. Consider the following option replication strategy for a call option with strike 30 on an underlying asset with spot price \$25: If the price of the asset goes above \$30, buy one unit of the asset for \$30 and hold it while the price is above \$30. If the price of the asset goes back below \$30, sell the one unit of the asset for \$30. Thus, at maturity, you will either hold no position, if the price of the asset is below the strike price 30 or you will have one unit of the asset which you bought for \$30, corresponding to the payoff of a call option with strike \$30. Seemingly, you replicated the call option at no cost. What is wrong with this argument?

27. An at–the–money straddle is made of a long at–the–money call and a long at–the–money put, both with strike equal to the spot price of the underlying asset.

    Consider a three months at–the–money straddle on an underlying asset with 20% volatility. Assuming that the risk–free rate and the dividend yield are zero, what is the approximate value of the delta of the straddle?

28. (i) What is the Put–Call parity for asset–or–nothing options?

    (ii) What is the Put–Call parity for cash–or–nothing options?

29. Why is the price of a plain vanilla option as a function of the strike of the option a strictly convex function?

30. What is the sign of the ρ of a plain vanilla European option and what is the intuition behind this?

31. Which one is more valuable, a three–months put option with strike \$90, or a three–months call option with strike \$110, on an underlying asset with spot price \$100?

2.4 C++. Data structures.

1. How do you declare an array?

2. How do you get the address of a variable?

3. How do you declare an array of pointers?

4. How do you declare a const pointer, a pointer to a const and a const pointer to a const?

5. How do you declare a dynamic array?

6. What is the general form for a function signature?

7. How do you pass-by-reference?

8. How do you pass a read only argument by reference?

9. What are the important differences between using a pointer and a reference?

10. How do you set a default value for a parameter?

11. How do you create a template function?

12. How do you declare a pointer to a function?

13. How do you prevent the compiler from doing an implicit conversion with your class?

14. Describe all the uses of the keyword `static` in C++.

15. Can a static member function be const?

16. C++ constructors support the initialization of member data via an initializer list. When is this preferable to initialization inside the body of the constructor?

17. What is a copy constructor, and how can the default copy constructor cause problems when you have pointer data members?

18. What is the output of the following code:

```
#include <iostream>
using namespace std;

class A
{
public:
    int * ptr;
    ~A()
    {
        delete(ptr);
    }
};
```

```
void foo(A object_input)
{
    ;
}

int main()
{
    A aa;
    aa.ptr = new int(2);
    foo(aa);
    cout<<(*aa.ptr)<<endl;
    return 0;
}
```

19. How do you overload an operator?

20. What are smart pointers?

21. What is encapsulation?

22. What is a polymorphism?

23. What is inheritance?

24. What is a virtual function? What is a pure virtual function and when do you use it?

25. Why are virtual functions use for destructors? Can they be used for constructors?

26. Write a function that computes the factorial of a positive integer.

27. Write a function that takes an array and returns the subarray with the largest sum.

28. Write a function that returns the prime factors of a positive integer.

29. Write a function that takes a 64-bit integer and swaps the bits at indices `i` and `j`.

30. Write a function that reverses a single linked list.

31. Write a function that takes a string and returns true if its parenthesis are balanced.

32. Write a function that returns the height of an arbitrary binary tree.

33. Write a C++ function that computes the n-th Fibonacci number.

34. Implement a basic calculator to evaluate a simple expression string. The expression string may contain open parentheses "(" and closing parentheses ")", the plus sign "+" or the minus sign "-", non-negative integers and empty spaces.

Note: You may assume that the given expression is always valid. Do not use the "eval" built-in library function.

Example 1:

```
Input: "1 + 1"
Output: 2
```

Example 2:

```
Input: "2-1 + 2"
Output: 3
```

Example 3:

```
Input: "(1+(4+5+2)-3)+(6+8)"
Output: 23
```

2.5 Statistics. Machine Learning.

1. Let X_i be i.i.d uniform random variables on $[-1, 1]$. Define the random variable

$$X = \frac{\sum_{i=1}^{n} X_i}{\sqrt{\sum_{i=1}^{n} X_i^2}}.$$

Calculate the probability of $X > 1$, as $n \to \infty$.

2. You are given a set of three standard 6-sided dice A, B, and C. The following table describes the outcomes of six rolls of these dices.

A	B	C
4	3	5
4	6	5
2	3	2
4	3	1
1	4	1
1	3	1

According to the data, which of them is the "most rigged"? Why?

3. You have an unfair coin which lands on heads 60% of the time. How many coin tosses are needed to detect that the coin is unfair?

4. Let $X_1, X_2, \ldots, X_n$ be an independent, identically distributed samples from the Gamma distribution $\Gamma\left(1, \frac{1}{\lambda}\right)$.

(a) Find the probability distribution function of the first order statistic $Y_1 = X_{(1)}$.

(b) What constant c makes cY_1 an unbiased estimator for λ?

(c) Find the MLE for λ. What is the *exact* sampling distribution of the MLE $\hat{\lambda}$? What is the bias of $\hat{\lambda}$? Find the *exact* variance of $\hat{\lambda}$.

(d) Find the Fisher information and state the asymptotic variance property for the MLE $\hat{\lambda}$ from the part (c). Compare the unbiased estimators found in parts (b) and (c). Is there any other unbiased estimator of λ better than $\hat{\lambda}$?

5. The response variable Y is regressed onto the regressor X, and the result is $Y = aX + b$. Then X and Y switch roles; that is, X as the response is regressed onto the Y as the regressor, and the result is $X = a'Y + b'$. What is the relationship between a and a'? What is the relationship between the R^2, t- and F-statistics of the two regressions?

6. Run the OLS regression of Y onto X_1; say R^2 is 0.4. Then, run the OLS regression of Y onto X_2; say R^2 is 0.5. Next, run the OLS regression of Y onto X_1 *and* X_2. What is the range of possible values for R^2 now?

7. Compute the Kendall's τ for the Clayton copula.

8. What are the differences between ordinary least square (OLS), ridge, LASSO, and ElasticNet regressions? How would you choose among them which method to use?

9. What is the bias-variance trade-off and how do you handle it?

10. How do you test the multicollinearity of a given set of data? How do you address the issue of significant multicollinearity?

11. What is the loss function employed in support vector regression (SVR)?

2.6 Monte Carlo simulations. Numerical methods.

1. How would you compute π using Monte Carlo simulations? What is the standard deviation of this method?

2. What methods do you know for generating independent samples of the standard normal distribution?

3. How do you generate a geometric Brownian motion stock path using random numbers from a normal distribution?

4. How do you generate a sample of the standard normal distribution from 12 independent samples of the uniform distribution on $[0, 1]$?

5. What is the rate of convergence for Monte Carlo methods?

6. What variance reduction techniques do you know?

7. How do you generate samples of normal random variables with correlation ρ?

8. What is the order of convergence of the Newton's method?

9. Which finite difference method corresponds to trinomial trees?

10. What is the relationship between the LU and Cholesky decompositions?

11. (i) Which matrices have an LU decomposition without pivoting?

    (ii) Does a symmetric positive definite matrix have an LU decomposition without pivoting?

2.7 Probability. Stochastic calculus.

1. What is the exponential distribution? What are the mean and the variance of the exponential distribution?

2. If X and Y are independent exponential random variables with mean 6 and 8, respectively, what is the probability that Y is greater than X?

3. What are the expected value and the variance of the Poisson distribution?

4. A point is chosen uniformly from the unit disk. What is the expected value of the distance between the point and the center of the disk?

5. Consider two random variables X and Y with mean 0 and variance 1, and with joint normal distribution. If $\mathrm{cov}(X,Y) = \frac{1}{\sqrt{2}}$, what is the conditional probability $P(X > 0|Y < 0)$?

6. If X and Y are lognormal random variables, is their product XY lognormally distributed?

7. Let X be a normal random variable with mean μ and variance σ^2, and let Φ be the cumulative distribution function of the standard normal distribution. Find the expected value of $Y = \Phi(X)$.

8. What is the law of large numbers?

9. What is the central limit theorem?

10. What is a martingale? How is it related to option pricing?

11. Explain the assumption $(dW_t)^2 = dt$ used in the informal derivation of Itô's Lemma.

12. If W_t is a Wiener process, find $E[W_t W_s]$.

13. If W_t is a Wiener process, what is $\mathrm{var}(W_t + W_s)$?

14. Let W_t be a Wiener process. Find

$$\int_0^t W_s \, dW_s \quad \text{and} \quad E\left[\int_0^t W_s \, dW_s\right].$$

15. Find the distribution of the random variable

$$X = \int_0^1 W_t dW_t.$$

16. Let W_t be a Wiener process. Find the mean and the variance of

$$\int_0^t W_s^2 dW_s.$$

17. If W_t is a Wiener process, find the variance of

$$X = \int_0^1 \sqrt{t} e^{\frac{W_t^2}{8}} dW_t.$$

18. If W_t is a Wiener process, what is $E\left[e^{W_t}\right]$?

19. If W_t is a Wiener process, find the variance of

$$\int_0^t s\, dW_s.$$

20. Let W_t be a Wiener process, and let

$$X_t = \int_0^t W_\tau d\tau.$$

What is the distribution of X_t? Is X_t a martingale?[2]

21. What is an Itô process?

22. What is Itô's lemma?

23. If W_t is a Wiener process, is the process $X_t = W_t^2$ a martingale?

[2] A solution to this question was given in Chapter 1 using integration by parts; a different solution will be given herein.

24. If W_t is a Wiener process, is the process

$$N_t = W_t^3 - 3tW_t$$

a martingale?

25. What is Girsanov's theorem?

26. What is the martingale representation theorem, and how is it related to option pricing and hedging?

27. Solve $dY_t = Y_t \, dW_t$, where W_t is a Wiener process.

28. Solve the following SDEs:

    (i) $dY_t = \mu Y_t dt + \sigma Y_t dW_t$;

    (ii) $dX_t = \mu dt + (aX_t + b)dW_t$.

29. What is the Heston model?

30. Show that the probability density function of the standard normal integrates to 1.

31. Let $W_t = (X_t, Y_t)$ be a two dimensional Brownian motion starting at (x, y), i.e., X_t and Y_t are independent one dimensional Brownian motions with $X_0 = x$ and $Y_0 = y$.

    (i) Find the probability that the Brownian motion W_t reaches the y-axis before reaching the x-axis.

(ii) Let $0 < r_1 < r_2$ such that $r_1 < \sqrt{x^2 + y^2} < r_2$. Find the probability that W_t enters the inner circle of center 0 and radius r_1 before leaving the outer circle of center 0 and radius r_2.

32. Let B_t, $t \geq 0$, be a standard Brownian motion in the probability measure $\mathbb{P}$. Determine the probability density function of $B_{\frac{3}{2}}$ under the probability measure $\tilde{\mathbb{P}}$ defined by the Radon–Nikodym derivative
$$\frac{d\tilde{\mathbb{P}}}{d\mathbb{P}} = e^{B_1 - \frac{1}{2}}.$$

33. Let $\boldsymbol{B}_t = (B_t^1, B_t^2)$ be a two dimensional Brownian motion in the (x, y) plane. Let $a > 0$ and denote by τ the first time $\boldsymbol{B}_t$ hits the line $y = a$. Determine the probability distribution of B_τ^1.

34. If Z is a standard normal variable, find the covariance of Z and Z^2.

35. If the function
$$f(x) = \frac{a}{1 + x^2},$$
where a is a real constant, is the probability density function of a random variable, find the value of a.

36. How many independent random variables uniformly distributed on $[0, 1]$ should you generate to ensure that there is at least one between 0.70 and 0.72 with probability 95%?

2.8 Brainteasers.

1. A flea is going between two points which are 100 inches apart by jumping (always in the same direction) either one inch or two inches at a time. How many different paths can the flea travel by?

2. I have a bag containing three pancakes: one golden on both sides, one burnt on both sides, and one golden on one side and burnt on the other. You shake the bag, draw a pancake at random, look at one side, and notice that it is golden. What is the probability that the other side is golden?

3. Alice and Bob are playing heads and tails, Alice tosses $n + 1$ coins, Bob tosses n coins. The coins are fair. What is the probability that Alice will have strictly more heads than Bob?

4. Alice is in a restaurant trying to decide between three desserts. How can she choose one of three desserts with equal probability with the help of a fair coin? What if the coin is biased and the bias is unknown?

5. What is the expected number of times you must flip a fair coin until it lands on head? What if the coin is biased and lands on head with probability p?

6. What is the expected number of coin tosses of a fair coin in order to get two heads in a row? What if

the coin is biased with 25% probability of getting heads?

7. A fair coin is tossed n times. What is the probability that no two consecutive heads appear?

8. You have two identical Fabergé eggs, either of which would break if dropped from the top of a building with 100 floors. Your task is to determine the highest floor from which an egg could be dropped without breaking. What is the minimum number of drops required to achieve this? You are allowed to break both eggs in the process.

9. An ant is in the corner of a $10 \times 10 \times 10$ room and wants to go to the opposite corner. What is the length of the shortest path the ant can take?

10. A $10 \times 10 \times 10$ cube is made of $1,000$ unit cubes. How many unit cubes can you see on the outside?

11. Fox Mulder is imprisoned by aliens in a large circular field surrounded by a fence. Outside the fence is a vicious alien that can run four times as fast as Mulder, but is constrained to stay near the fence. If Mulder can contrive to get to an unguarded point on the fence, he can quickly scale the fence and escape. Can he get to a point on the fence ahead of the alien?

12. At your subway station, you notice that of the two trains running in opposite directions which are supposed to arrive with the same frequency, the train

going in one direction comes first 80% of the time, while the train going in the opposite direction comes first only 20% of the time. What do you think could be happening?

13. You start off with one amoeba. Every minute, this amoeba can either die, do nothing, split into two amoebas, or split into three amoebas; all these scenarios being equally likely to happen. All further amoebas behave the same way. What is the probability that the amoebas eventually die off?

14. Given a set X with n elements, choose two subsets A and B at random. What is the probability of A being a subset of B?

15. Alice writes two distinct real numbers between 0 and 1 on two sheets of paper. Bob selects one of the sheets randomly to inspect it. He then has to declare whether the number he sees is the bigger or smaller of the two. Is there any way Bob can expect to be correct more than half the times Alice plays this game with him?

16. How many digits does the number 125^{100} have? You are not allowed to use values of $\log_{10} 2$ or $\log_{10} 5$.

17. For every subset of $\{1, 2, 3, \ldots, 2013\}$, arrange the numbers in the increasing order and take the sum with alternating signs. The resulting integer is called

the weight of the subset.[3] Find the sum of the weights of all the subsets of $\{1, 2, 3, \ldots, 2013\}$.

18. Alice and Bob alternately choose one number from one of the following nine numbers: 1/16, 1/8, 1/4, 1/2, 1, 2, 4, 8, 16, without replacement. Whoever gets three numbers that multiply to one wins the game. Alice starts first. What should her strategy be? Can she always win?

19. Mr. and Mrs. Jones invite four other couples over for a party. At the end of the party, Mr. Jones asks everyone else how many people they shook hands with, and finds that everyone gives a different answer. Of course, no one shook hands with his or her spouse and no one shook the same person's hand twice. How many people did Mrs. Jones shake hands with?

20. The New York Yankees and the San Francisco Giants are playing in the World Series (best of seven format). You would like to bet \$100 on the Yankees winning the World Series, but you can only place bets on individual games, and every time at even odds. How much should you bet on the first game?

21. We have two red, two green and two yellow balls. For each color, one ball is heavy and the other is light. All heavy balls weigh the same. All light balls weigh the same. How many weighings on a scale are necessary to identify the three heavy balls?

[3] For example, the weight of the subset $\{3\}$ is 3. The weight of the subset $\{2, 5, 8\}$ is $2 - 5 + 8 = 5$.

22. There is a row of 10 rooms and a treasure in one of them. Each night, a ghost moves the treasure to an adjacent room. You are trying to find the treasure, but can only check one room per day. How do you find it?

23. How many comparisons do you need to find the maximum in a set of n distinct numbers? How many comparisons do you need to find both the maximum and minimum in a set of n distinct numbers?

24. Given a cube, you can jump from one vertex to a neighboring vertex with equal probability. Assume you start from a certain vertex (does not matter which one). What is the expected number of jumps to reach the opposite vertex?

25. Select numbers uniformly distributed between 0 and 1, one after the other, as long as they keep decreasing; i.e. stop selecting when you obtain a number that is greater than the previous one you selected.

    (i) On average, how many numbers have you selected?

    (ii) What is the average value of the smallest number you have selected?

26. To organize a charity event that costs \$100K, an organization raises funds. Independent of each other, one donor after another donates some amount of money that is exponentially distributed with a mean of \$20K. The process is stopped as soon as \$100K or more has been collected. Find the distribution,

mean, and variance of the number of donors needed until at least \$100K has been collected.

27. Consider a random walk starting at 1 and with equal probability of moving to the left or to the right by one unit, and stopping either at 0 or at 3.

    (i) What is the expected number of steps to do so?

    (ii) What is the probability of the random walk ending at 3 rather than at 0?

28. A stick of length 1 drops and breaks at a random place uniformly distributed across the length. What is the expected length of the smaller part?

29. You are given a stick of unit length.

    (i) The stick drops and breaks at two places. What is the probability that the three pieces could form a triangle?

    (ii) The stick drops and breaks at one place. Then the larger piece is taken and dropped again, breaking at one place. What is the probability that the three pieces could form a triangle?

30. Why is a manhole cover round?

31. When is the first time after 12 o'clock that the hour and minute hands of a clock meet again?

32. Three light switches are in one room, and they turn three light bulbs in another. How do you figure out which switch turns on which bulb in one shot?

33. The number 2^{29} has 9 digits, all different. Without computing 2^{29}, find the missing digit.

34. Alice and Bob stand at opposite ends of a straight line segment. Bob sends 50 ants towards Alice, one after another. Alice sends 20 ants towards Bob. All ants travel along the straight line segment. Whenever two ants collide, they simply bounce back and start traveling in the opposite direction. How many ants reach Bob and how many ants reach Alice? How many ant collisions take place?

35. There are 20 people at a party. Everyone writes down their name on a piece of paper and throws it in a bag. We shake up the bag and each person draws one name from the bag. You are in the same group as the person you have drawn.

    For example, if people labeled 1 through 20 drew the following names from the bag:

$$\begin{array}{cccccccccc} 1 & 2 & 3 & 4 & 5 & 6 & 7 & 8 & 9 & 10 \\ \downarrow & \downarrow & \downarrow & \downarrow & \downarrow & \downarrow & \downarrow & \downarrow & \downarrow & \downarrow \\ 6 & 5 & 3 & 8 & 11 & 10 & 9 & 13 & 15 & 12 \end{array}$$

$$\begin{array}{cccccccccc} 11 & 12 & 13 & 14 & 15 & 16 & 17 & 18 & 19 & 20 \\ \downarrow & \downarrow & \downarrow & \downarrow & \downarrow & \downarrow & \downarrow & \downarrow & \downarrow & \downarrow \\ 14 & 18 & 4 & 20 & 17 & 16 & 7 & 19 & 1 & 2 \end{array}$$

    the groups that form are

$$(1, 6, 10, 12, 18, 19), (2, 5, 11, 14, 20), (3),$$

$$(4, 8, 13), (7, 9, 15, 17), (16).$$

What is the expected number of groups?

36. Let A be the sum of the digits of 2019!. Let B be the sum of the digits of A. Let C be the sum of the digits of B. Find C.

37. Find 2019 consecutive positive integers that are not prime.

38. Exactly 4 out of 100 coins are fake. All the genuine coins weigh the same; all the fake coins, too. A fake coin is lighter than a genuine coin. How can you find at least one genuine coin using a balance scale only twice?

39. Can you design a pair of 6–sided non–identical fair dice different from the standard dice with each face bearing a positive integer and having the same probability distribution for the sum as the pair of standard dice? (In other words, there must be two ways to roll a 3, six ways to roll a 7, one way to roll a 12, and so forth.)

40. In the two–dimensional space, for every equilateral triangle ABC, we have that

$$f(A) + f(B) + f(C) = 0.$$

What can you say about the function f?

41. Suppose you have in your possession an incredibly large bag of M&M's containing a uniform distribution of the six M&M colors. (M&M's come in blue, orange, green, yellow, red, brown.) You decide to play a game: you draw one M&M from the bag and place it on the table. You then continue to draw M&M's from the bag one at a time. If you draw an M&M that is the same color as one already on the table, you eat both of them. Otherwise, you place the M&M on the table along with the others of different color. The game ends when you have six M&M's (all of different colors) on the table. How many M&M's should you expect to eat playing this game?

Chapter 3

Solutions

3.1 Mathematics, calculus, differential equations.

Question 1. What is the value of i^i, where $i = \sqrt{-1}$?

Answer: Recall that $e^{i\theta} = \cos\theta + i\sin\theta$. Then,

$$i = \cos\frac{\pi}{2} + i\sin\frac{\pi}{2} = e^{i\frac{\pi}{2}},$$

and therefore

$$i^i = \left(e^{i\frac{\pi}{2}}\right)^i = e^{(i\frac{\pi}{2})i} = e^{i^2\frac{\pi}{2}} = e^{-\frac{\pi}{2}},$$

since $i^2 = -1$. □

Question 2. Which number is larger, π^e or e^π?

Answer: We will show that $\pi^e < e^\pi$. By taking the natural logarithm, we find that

$$\begin{aligned} \pi^e < e^\pi \quad &\Longleftrightarrow \quad \ln(\pi^e) < \ln(e^\pi) \quad \Longleftrightarrow \quad e\ln(\pi) < \pi \\ &\Longleftrightarrow \quad \frac{\ln(\pi)}{\pi} < \frac{1}{e}, \end{aligned} \tag{3.1}$$

which can be written as

$$\frac{\ln(\pi)}{\pi} < \frac{\ln(e)}{e}.$$

Let $f : (0, \infty) \to \mathbb{R}$ given by $f(x) = \frac{\ln(x)}{x}$. Then,

$$f'(x) = \frac{1 - \ln(x)}{x^2}.$$

Note that $f'(x) = 0$ has one solution, $x = e$. Also, $f'(x) > 0$ for $0 < x < e$, and $f'(x) < 0$ for $x > e$, and therefore $f(x)$ is increasing on the interval $(0, e)$ and is decreasing on the interval (e, ∞).

Thus, the function $f(x) = \frac{\ln(x)}{x}$ has a global maximum point at $x = e$, i.e., $f(x) < f(e) = \frac{1}{e}$ for all $x > 0$ with $x \neq e$, and therefore

$$f(\pi) = \frac{\ln(\pi)}{\pi} < \frac{\ln(e)}{e} = \frac{1}{e},$$

which is equivalent to $\pi^e < e^\pi$; cf. (3.1). □

Question 3. Show that

$$\frac{e^x + e^y}{2} \geq e^{\frac{x+y}{2}}, \quad \forall\, x, y \in \mathbb{R}. \tag{3.2}$$

Answer: Let $e^x = a$ and $e^y = b$. Note that $a, b > 0$, and that

$$e^{\frac{x+y}{2}} = \sqrt{e^{x+y}} = \sqrt{e^x \cdot e^y} = \sqrt{ab}.$$

Then, (3.2) can be written as

$$\frac{a+b}{2} \geq \sqrt{ab} \iff a + b - 2\sqrt{ab} \geq 0$$
$$\iff \left(\sqrt{a} - \sqrt{b}\right)^2 \geq 0,$$

which is what we wanted to show. □

Question 4. Solve $x^6 = 64$.

Answer: Recall that the six unit roots of $z^6 = 1$ are

$$\begin{aligned} z_k &= \exp\left(\frac{2k\pi i}{6}\right) = \exp\left(\frac{k\pi i}{3}\right) \\ &= \cos\left(\frac{k\pi}{3}\right) + i\sin\left(\frac{k\pi}{3}\right), \end{aligned} \tag{3.3}$$

for $k = 0 : 5$. Since $\sqrt[6]{64} = 2$, we obtain from (3.3) that the solutions of $x^6 = 64$ are

$$x_k = 2\cos\left(\frac{k\pi}{3}\right) + 2i\sin\left(\frac{k\pi}{3}\right), \quad \forall\, k = 0 : 5. \quad \square$$

Question 5. What is the derivative of x^x?

Answer: Note that

$$x^x = e^{\ln(x^x)} = e^{x\ln(x)}. \tag{3.4}$$

Using Chain Rule and (3.4), we find that

$$\begin{aligned} (x^x)' &= \left(e^{x\ln(x)}\right)' = e^{x\ln(x)}\,(x\ln(x))' \\ &= x^x(\ln(x) + 1). \quad \square \end{aligned}$$

Question 6. Calculate

$$\sqrt{2 + \sqrt{2 + \sqrt{2 + \ldots}}}. \tag{3.5}$$

Answer: Assume that the limit from (3.5) exists, and denote that limit by l. Then, $l = \sqrt{2 + l}$, which can be written as

$$l^2 - l - 2 = (l-2)(l+1) = 0.$$

Since $l > 0$, we obtain that $l = 2$, i.e.,

$$\sqrt{2 + \sqrt{2 + \sqrt{2 + \dots}}} = 2. \tag{3.6}$$

Thus, proving (3.6) is equivalent to showing that the sequence $(x_n)_{n \geq 0}$ given by $x_0 = \sqrt{2}$ and

$$x_{n+1} = \sqrt{2 + x_n}, \quad \forall\, n \geq 0,$$

is convergent.

We can see by induction that the sequence $(x_n)_{n \geq 0}$ is bounded from above by 2, since $x_0 = \sqrt{2} < 2$, and, if we assume that $x_n < 2$, then

$$x_{n+1} = \sqrt{2 + x_n} < \sqrt{4} = 2.$$

Moreover, the sequence $(x_n)_{n \geq 0}$ is increasing, since

$$\begin{aligned} x_n < x_{n+1} &\iff x_n < \sqrt{2 + x_n} \\ &\iff x_n^2 - x_n - 2 < 0 \\ &\iff (x_n - 2)(x_n + 1) < 0, \end{aligned}$$

which holds true since $x_n > 0$, and since, as shown above, $x_n < 2$ for all $n \geq 0$.

Thus, the sequence $(x_n)_{n \geq 0}$ is convergent since it is increasing and bounded from above, which is what we needed to show in order to prove (3.6). $\square$

Question 7. Find x such that

$$x^{x^{x^{\cdot^{\cdot^{\cdot}}}}} = 2. \tag{3.7}$$

Answer: If x exists such that (3.7) holds true, then

$$x^{x^{x^{x^{\cdot^{\cdot^{\cdot}}}}}} = x^2 = 2,$$

and therefore the only possible solution to (3.7) is $x = \sqrt{2}$. We prove that $x = \sqrt{2}$ is, indeed, the solution to (3.7), by showing that

$$\sqrt{2}^{\sqrt{2}^{\sqrt{2}^{\cdot^{\cdot}}}} = 2. \tag{3.8}$$

Consider the sequence $(x_n)_{n\geq 0}$ with $x_0 = \sqrt{2}$ and satisfying the following recursion:

$$x_{n+1} = \left(\sqrt{2}\right)^{x_n} = 2^{x_n/2}, \quad \forall\, n \geq 0. \tag{3.9}$$

We can see by induction that $(x_n)_{n\geq 0}$ is an increasing sequence, since $x_0 = \sqrt{2} < \sqrt{2}^{\sqrt{2}} = x_1$, and, if we assume that $x_{n-1} < x_n$, then

$$x_n = 2^{x_{n-1}/2} < 2^{x_n/2} = x_{n+1}.$$

Also by induction, we can prove that the sequence $(x_n)_{n\geq 0}$ is bounded from above by 2, since $x_0 = \sqrt{2} < 2$, and, if $x_n < 2$, then

$$x_{n+1} = 2^{x_n/2} < 2.$$

Thus, the sequence $(x_n)_{n\geq 0}$ is convergent since it is increasing and bounded from above.

Let $l = \lim_{n\to\infty} x_n$. From (3.9), we find that $l = 2^{l/2}$, which is equivalent to

$$l^{1/l} = 2^{1/2}. \tag{3.10}$$

The function $f : (0,\infty) \to (0,\infty)$ given by

$$\begin{aligned} f(t) &= t^{1/t} = \exp\left(\ln(t^{1/t})\right) \\ &= \exp\left(\frac{\ln(t)}{t}\right) \end{aligned}$$

is increasing for $t < e$ and decreasing for $t > e$, since

$$f'(t) = \frac{1 - \ln(t)}{t^2} \exp\left(\frac{\ln(t)}{t}\right)$$

and $f'(t) > 0$ for $t < e$ and $f'(t) < 0$ for $t > e$.

Thus, there are two values of l such that (3.10) is satisfied, i.e., such that $l^{1/l} = 2^{1/2}$, one value being equal to 2, and the other one greater than e. Since we showed that $x_n < 2$ for all $n \geq 0$, we conclude that $l = \lim_{n\to\infty} x_n = 2$, which is what we needed to show to complete the proof of (3.8). □

Question 8. Which of the following series converge:

$$\sum_{k=1}^{\infty} \frac{1}{k}; \quad \sum_{k=1}^{\infty} \frac{1}{k^2}; \quad \sum_{k=2}^{\infty} \frac{1}{k\ln(k)}?$$

Answer: We show that

$$\sum_{k=1}^{\infty} \frac{1}{k^2} \quad \text{is convergent;}$$

$$\sum_{k=1}^{\infty} \frac{1}{k} \quad \text{and} \quad \sum_{k=2}^{\infty} \frac{1}{k\ln(k)} \quad \text{are divergent.}$$

Since all the terms of the series $\sum_{k=1}^{\infty} \frac{1}{k^2}$ are positive, it is enough to show that the partial sums $\sum_{k=1}^{n} \frac{1}{k^2}$ are uniformly bounded, in order to conclude that the series is convergent. This can be seen as follows:

$$\begin{aligned}
\sum_{k=1}^{n} \frac{1}{k^2} &= 1 + \sum_{k=2}^{n} \frac{1}{k^2} \\
&\leq 1 + \sum_{k=2}^{n} \frac{1}{k(k-1)} \\
&= 1 + \sum_{k=2}^{n} \left(\frac{1}{k-1} - \frac{1}{k}\right) \\
&= 1 + \left(1 - \frac{1}{n}\right) \\
&< 2, \quad \forall\ n \geq 2.
\end{aligned}$$

To show that the series $\sum_{k=1}^{\infty} \frac{1}{k}$ is divergent, we will prove that

$$\sum_{k=1}^{n} \frac{1}{k} > \ln(n) + \frac{1}{n}, \quad \forall\, n \geq 1. \tag{3.11}$$

Since $\frac{1}{x}$ is a decreasing function, it follows that

$$\frac{1}{x} < \frac{1}{k}, \quad \forall\, k < x < k+1.$$

Then,

$$\begin{aligned} \int_1^n \frac{1}{x}\, dx &= \sum_{k=1}^{n-1} \int_k^{k+1} \frac{1}{x}\, dx \\ &< \sum_{k=1}^{n-1} \int_k^{k+1} \frac{1}{k}\, dx \\ &= \sum_{k=1}^{n-1} \frac{1}{k} \\ &= -\frac{1}{n} + \sum_{k=1}^{n} \frac{1}{k}. \end{aligned} \tag{3.12}$$

From (3.12), we find that

$$\begin{aligned} \sum_{k=1}^{n} \frac{1}{k} &> \int_1^n \frac{1}{x}\, dx + \frac{1}{n} \\ &= \ln(n) + \frac{1}{n}, \quad \forall\, n \geq 1, \end{aligned}$$

which is what we wanted to prove; see (3.11).

Similarly, note that

$$\frac{1}{x\ln(x)} < \frac{1}{k\ln(k)}, \quad \forall\, k < x < k+1,$$

and therefore

$$\begin{aligned} \int_2^{n+1} \frac{1}{x\ln(x)}\, dx &= \sum_{k=2}^{n} \int_k^{k+1} \frac{1}{x\ln(x)} dx \\ &< \sum_{k=2}^{n} \int_k^{k+1} \frac{1}{k\ln(k)} dx \\ &= \sum_{k=2}^{n} \frac{1}{k\ln(k)}. \end{aligned} \tag{3.13}$$

Since

$$\int_2^{n+1} \frac{1}{x\ln(x)}\, dx = \ln(\ln(n+1)) - \ln(\ln(2)),$$

we obtain from (3.13) that

$$\sum_{k=2}^{n} \frac{1}{k\ln(k)} > \ln(\ln(n+1)) - \ln(\ln(2)),$$

and we conclude that the series $\sum_{k=2}^{\infty} \frac{1}{k\ln(k)}$ is divergent. Note: Although not needed to answer this question, it can be shown that

$$\sum_{k=1}^{\infty} \frac{1}{k^2} = \frac{\pi^2}{6},$$

and

$$\lim_{n\to\infty} \left(\sum_{k=1}^{\infty} \frac{1}{k} - \ln(n) \right) = \gamma,$$

where $\gamma \approx 0.57721$ is Euler's constant. □

Question 9. Compute

$$\int \frac{1}{1+x^2}\, dx.$$

Answer: Use the substitution $x = \tan(z)$. Then, $dx = \frac{1}{\cos^2(z)} dz$ and

$$\begin{aligned}\int \frac{1}{1+x^2}\, dx &= \int \frac{1}{(1+\tan^2(z))\cos^2(z)}\, dz \\ &= \int \frac{1}{\cos^2(z)+\sin^2(z)}\, dz \\ &= \int 1\, dz \\ &= z + C,\end{aligned}$$

where C is a real constant, since $\cos^2(z) + \sin^2(z) = 1$ for any z. Solving $x = \tan(z)$ for z, we obtain that $z = \arctan(x)$, and therefore

$$\int \frac{1}{1+x^2}\, dx = \arctan(x) + C. \quad \square$$

Question 10. Compute

$$\int x\ln(x)dx \quad \text{and} \quad \int xe^x dx.$$

Answer: By integration by parts,

$$\begin{aligned}\int x\ln(x)\, dx &= \frac{x^2}{2}\ln(x) - \int \frac{x^2}{2}\cdot\frac{1}{x}\, dx \\ &= \frac{x^2}{2}\ln(x) - \frac{1}{2}\int x\, dx \\ &= \frac{x^2\ln(x)}{2} - \frac{x^2}{4} + C; \\ \int xe^x\, dx &= xe^x - \int 1\cdot e^x\, dx \\ &= xe^x - e^x + C. \quad \square\end{aligned}$$

Question 11. Compute

$$\int x^n \ln(x)\, dx.$$

Answer: If $n \neq -1$, we use integration by parts and find that

$$\begin{aligned} \int x^n \ln(x)\, dx &= \frac{x^{n+1}}{n+1}\ln(x) - \int \frac{x^{n+1}}{n+1}\cdot\frac{1}{x}\, dx \\ &= \frac{x^{n+1}\ln(x)}{n+1} - \frac{1}{n+1}\int x^n\, dx \\ &= \frac{x^{n+1}\ln(x)}{n+1} - \frac{x^{n+1}}{(n+1)^2} + C. \end{aligned}$$

For $n = -1$, we obtain that

$$\int \frac{\ln(x)}{x}\, dx = \frac{(\ln(x))^2}{2} + C,$$

where C is a real constant, since

$$\left(\frac{(\ln(x))^2}{2}\right)' = \frac{1}{2}\cdot 2\ln(x)\cdot(\ln(x))' = \frac{\ln(x)}{x}. \quad \square$$

Question 12. Compute

$$\int (\ln(x))^n\, dx.$$

Answer: For every integer $n \geq 0$, let

$$f_n(x) = \int (\ln(x))^n\, dx.$$

By using integration by parts, we find that, for any $n \geq 1$,

$$\begin{aligned}
& \int (\ln(x))^n \, dx \\
= \; & x(\ln(x))^n \; - \; \int x\left((\ln(x))^n\right)' \, dx \\
= \; & x(\ln(x))^n \; - \; \int x \cdot n(\ln(x))^{n-1} \cdot (\ln(x))' \, dx \\
= \; & x(\ln(x))^n \; - \; \int x \cdot n(\ln(x))^{n-1} \cdot \frac{1}{x} \, dx \\
= \; & x(\ln(x))^n - n\int (\ln(x))^{n-1} \, dx,
\end{aligned}$$

and therefore

$$f_n(x) \;=\; x(\ln(x))^n - nf_{n-1}(x), \quad \forall\, n \geq 1. \tag{3.14}$$

Note that

$$f_0(x) \;=\; \int 1dx \;=\; x + C.$$

Thus, the recursion (3.14) can be used to find the values of $f_n(x)$ for all n. For example:

$$\begin{aligned}
f_1(x) &= x\ln(x) - f_0(x) \;=\; x(\ln(x) - 1) + C; \\
f_2(x) &= x(\ln(x))^2 - 2f_1(x) \\
&= x\left((\ln(x))^2 - 2\ln(x) + 2\right) + C.
\end{aligned}$$

The following general formula can be obtained by induction:

$$\int (\ln(x))^n \, dx \;=\; x\sum_{k=0}^{n} \frac{(-1)^{n-k} n!}{k!} (\ln(x))^k \;+\; C,$$

for all $n \geq 0$. □

Question 13. Solve the ODE

$$y'' - 4y' + 4y = 1. \tag{3.15}$$

Answer: Note that (3.15) is a second order non homogeneous linear ODE with constant coefficients. The homogeneous ODE associated to (3.15) is

$$y'' - 4y' + 4y = 0, \tag{3.16}$$

whose characteristic equation, $z^2 - 4z + 4 = 0$, has a double root $z_1 = z_2 = 2$. Thus, the solution to the homogeneous ODE (3.16) is

$$y(x) \; = \; c_1 e^{2x} \; + \; c_2 x e^{2x}, \tag{3.17}$$

where c_1 and c_2 are constants.

Since the constant function $y_0(x) = \frac{1}{4}$ is a solution to the non homogeneous ODE (3.15), we conclude from (3.17) that the general form of the solution of (3.15) is

$$y(x) \; = \; c_1 e^{2x} \; + \; c_2 x e^{2x} \; + \; \frac{1}{4}. \quad \square$$

Question 14. Find $f(x)$ such that

$$f'(x) \; = \; f(x)(1 - f(x)). \tag{3.18}$$

Answer: Note that (3.18) is as an ODE with separable variables and can be written as follows:

$$\frac{y'}{y(1-y)} \; = \; 1, \tag{3.19}$$

where $y = f(x)$. By integrating (3.19) with respect to x we obtain that

$$\int \frac{y'}{y(1-y)} \, dx \; = \; \int 1 \, dx \; = \; x + C_1, \tag{3.20}$$

where $C_1 \in \mathbb{R}$ is a real constant.

Note that $dy = y'dx$, and therefore

$$\begin{aligned} \int \frac{y'}{y(1-y)}\, dx &= \int \frac{1}{y(1-y)}\, dy \\ &= \int \left(\frac{1}{y} + \frac{1}{1-y}\right) dy \\ &= \int \frac{1}{y}\, dy + \int \frac{1}{1-y}\, dy \\ &= \ln(|y|) - \ln(|1-y|) \\ &= \ln\left|\frac{y}{1-y}\right|. \end{aligned} \tag{3.21}$$

From (3.20) and (3.21), it follows that

$$\ln\left|\frac{y}{1-y}\right| = x + C_1,$$

and therefore

$$\left|\frac{y}{1-y}\right| = e^{x+C_1} = C_2 e^x,$$

where $C_2 = e^{C_1} > 0$ is a positive real constant.

Thus, either $\frac{y}{1-y} = C_2e^x$, or $\frac{y}{1-y} = -C_2e^x$, which can be written as

$$\frac{y}{1-y} = Ce^x, \tag{3.22}$$

where C is a real constant.

From (3.22), we obtain that $y = \frac{Ce^x}{1+Ce^x}$. We conclude that the ODE (3.18) has the following solution:

$$f(x) = \frac{Ce^x}{1+Ce^x},$$

where $C \in \mathbb{R}$ is a fixed constant. □

Question 15. Derive the Black-Scholes PDE.

Answer: Consider an asset with spot price S following a lognormal distribution with drift μ and volatility σ and paying dividends continuously at rate q. Then,

$$dS = (\mu - q)Sdt + \sigma S dW_t,$$

where W_t, $t \geq 0$, is a Wiener process.

Let $V = V(S,t)$ be the value at time t of a replicable non path dependent derivative security on this asset, when the underlying is priced at S. Set up a portfolio Π made of a long position in the derivative security V and a short position in

$$\Delta = \frac{\partial V}{\partial S} \tag{3.23}$$

units of the asset. Then,

$$\Pi = V - \Delta S.$$

Denote by dS, dV, and $d\Pi$ the changes in the values of S, V, and Π, respectively, over an infinitesimally small time period dt. Then,

$$d\Pi = dV - \Delta dS - \Delta qSdt, \tag{3.24}$$

where $\Delta qSdt$ is the dividend payment owed over the time dt on the short Δ units asset position. From (3.23) and (3.24), we find that

$$d\Pi = dV - \frac{\partial V}{\partial S}dS - qS\frac{\partial V}{\partial S}dt. \tag{3.25}$$

From Itô's formula, it follows that

$$\begin{aligned} dV &= \left(\frac{\partial V}{\partial t} + \frac{\sigma^2 S^2}{2}\frac{\partial^2 V}{\partial S^2}\right) dt \\ &+ \frac{\partial V}{\partial S}dS, \end{aligned} \tag{3.26}$$

and, from (3.25) and (3.26), we obtain that

$$d\Pi = \left(\frac{\partial V}{\partial t} + \frac{\sigma^2 S^2}{2}\frac{\partial^2 V}{\partial S^2} - qS\frac{\partial V}{\partial S}\right) dt, \tag{3.27}$$

which means that the value of the portfolio Π is deterministic over the small time period dt. For no–arbitrage, the value of the portfolio Π must grow at the risk–free rate over the time period dt, i.e., $d\Pi = r\Pi dt$, where r denotes the risk–free rate. Thus,

$$\begin{aligned} d\Pi &= r\Pi dt = r(V - \Delta S)dt \\ &= \left(rV - rS\frac{\partial V}{\partial S}\right) dt. \end{aligned} \tag{3.28}$$

From (3.27) and (3.28), we find that

$$\frac{\partial V}{\partial t} + \frac{\sigma^2 S^2}{2}\frac{\partial^2 V}{\partial S^2} - qS\frac{\partial V}{\partial S} = rV - rS\frac{\partial V}{\partial S},$$

and therefore

$$\frac{\partial V}{\partial t} + \frac{\sigma^2 S^2}{2}\frac{\partial^2 V}{\partial S^2} + (r - q)S\frac{\partial V}{\partial S} - rV = 0,$$

which is the Black-Scholes PDE for $V(S, t)$. □

Question 16. Calculate the limit

$$\lim_{x \to 0} \frac{(\cos(2x) - 1)^2 (\ln(1 + 3x) - 3x)}{x^3(e^{x^3} - 1)}$$

Answer: The limit could be calculated by applying l'Hôpital's rule six (!) times. However, this would be extremely laborious and not what an interviewer asking this question would want to see.

The approach that will lead to a quick solution is to use the Taylor approximations of the cosine, natural log, and exponential functions around zero. Recall that, as $x \to 0$,

$$\cos(x) \approx 1 - \frac{x^2}{2}; \tag{3.29}$$

$$\ln(1 + x) \approx x - \frac{x^2}{2}; \tag{3.30}$$

$$e^x \approx 1 + x. \tag{3.31}$$

Then,

$$\begin{aligned}
\cos(2x) &\overset{(3.29)}{\approx} 1 - \frac{(2x)^2}{2} = 1 - 2x^2; \\
\ln(1+3x) &\overset{(3.30)}{\approx} 3x - \frac{(3x)^2}{2} = 3x - \frac{9}{2}x^2; \\
e^{x^3} &\overset{(3.31)}{\approx} 1 + x^3,
\end{aligned}$$

and therefore

$$\cos(2x) - 1 \approx -2x^2; \tag{3.32}$$

$$\ln(1+3x) - 3x \approx -\frac{9}{2}x^2; \tag{3.33}$$

$$e^{x^3} - 1 \approx x^3, \tag{3.34}$$

as $x \to 0$.

From (3.32–3.34), we conclude that

$$\begin{aligned}
&\lim_{x\to 0} \frac{(\cos(2x)-1)^2(\ln(1+3x)-3x)}{x^3(e^{x^3}-1)} \\
= &\lim_{x\to 0} \frac{(-2x^2)^2 \cdot (-\frac{9}{2}x^2)}{x^3 \cdot x^3} \\
= &\lim_{x\to 0} \frac{4x^4 \cdot (-\frac{9}{2}x^2)}{x^6} \\
= &-18. \quad \square
\end{aligned}$$

Question 17. (i) Prove that

$$\sum_{i=1}^{n} i^3 = \left(\frac{n(n+1)}{2}\right)^2,$$

for all positive integers n.

(ii) Derive the formula

$$\sum_{i=1}^{n} i^3 = \left(\frac{n(n+1)}{2}\right)^2.$$

Answer: (i) We will use induction to prove that

$$\sum_{i=1}^{n} i^3 = \left(\frac{n(n+1)}{2}\right)^2, \tag{3.35}$$

for all positive integers n.

When $n = 1$, (3.35) becomes $1 = 1$ and therefore holds true.

Assume that (3.35) holds true for n, i.e., assume that

$$\sum_{i=1}^{n} i^3 = \left(\frac{n(n+1)}{2}\right)^2 = \frac{n^2(n+1)^2}{4}. \tag{3.36}$$

We will show that (3.35) also holds for $n + 1$, i.e.,

$$\sum_{i=1}^{n+1} i^3 = \left(\frac{(n+1)(n+2)}{2}\right)^2 = \frac{(n+1)^2(n+2)^2}{4}. \tag{3.37}$$

By using the induction hypothesis (3.36), the left hand side of (3.37) can be written as

$$\begin{aligned}
\sum_{i=1}^{n+1} i^3 &= \sum_{i=1}^{n} i^3 + (n+1)^3 \\
&\overset{(3.36)}{=} \frac{n^2(n+1)^2}{4} + (n+1)^3 \\
&= (n+1)^2\left(\frac{n^2}{4} + n + 1\right) \\
&= (n+1)^2 \cdot \frac{n^2 + 4n + 4}{4} \\
&= \frac{(n+1)^2(n+2)^2}{4},
\end{aligned}$$

which is the same as the right hand side of (3.37).

Thus, (3.37) holds true and the formula (3.35) is proved by induction.

(ii) The formulas for the sums of powers of consecutive are derived recursively. Thus, we assume that we already know the following formulas:

$$S_1 = \sum_{i=1}^{n} i = \frac{n(n+1)}{2}; \tag{3.38}$$

$$S_2 = \sum_{i=1}^{n} i^2 = \frac{n(n+1)(2n+1)}{6}, \tag{3.39}$$

and we will derive the formula

$$S_3 = \sum_{i=1}^{n} i^3 = \left(\frac{n(n+1)}{2}\right)^2. \tag{3.40}$$

Recall from the binomial formula that

$$(i+1)^4 = i^4 + 4i^3 + 6i^2 + 4i + 1, \tag{3.41}$$

for all $i = 1 : n$. We sum (3.41) over i from 1 to n and obtain that

$$\begin{aligned}
&\sum_{i=1}^{n}(i+1)^4 \\
=\ &\sum_{i=1}^{n}\left(i^4 + 4i^3 + 6i^2 + 4i + 1\right) \\
=\ &\sum_{i=1}^{n} i^4 + 4\sum_{i=1}^{n} i^3 + 6\sum_{i=1}^{n} i^2 + 4\sum_{i=1}^{n} i + \sum_{i=1}^{n} 1 \\
=\ &\sum_{i=1}^{n} i^4 + 4S_3 + 6S_2 + 4S_1 + n.
\end{aligned}$$

Thus,

$$4S_3 = \sum_{i=1}^{n}(i+1)^4 - \sum_{i=1}^{n} i^4 \tag{3.42}$$

$$- 6S_2 - 4S_1 - n. \tag{3.43}$$

Note that

$$\begin{aligned}
& \sum_{i=1}^{n}(i+1)^4 - \sum_{i=1}^{n} i^4 \\
= & \left((n+1)^4 + \sum_{i=2}^{n} i^4\right) - \left(1 + \sum_{i=2}^{n} i^4\right) \\
= & (n+1)^4 - 1. \qquad (3.44)
\end{aligned}$$

By substituting (3.44) in (3.42), and by substituting (3.38) and (3.39) in (3.43), we obtain that

$$\begin{aligned}
4S_3 &= (n+1)^4 - 1 \\
&\quad - 6 \cdot \frac{n(n+1)(2n+1)}{6} - 4 \cdot \frac{n(n+1)}{2} - n \\
&= n^4 + 4n^3 + 6n^2 + 4n \\
&\quad - (2n^3 + 3n^2 + n) - 2(n^2 + n) - n \\
&= n^4 + 2n^3 + n^2 \\
&= n^2(n+1)^2.
\end{aligned}$$

Thus,

$$S_3 = \frac{n^2(n+1)^2}{4} = \left(\frac{n(n+1)}{2}\right)^2,$$

and therefore the formula (3.40) is established. □

Question 18. Prove the Cauchy–Schwartz inequality

$$\left(\sum_{i=1}^{n} x_i^2\right)\left(\sum_{i=1}^{n} y_i^2\right) \geq \left(\sum_{i=1}^{n} x_i y_i^2\right)^2,$$

where x_i and y_i are real numbers, for $i = 1 : n$.

Answer: Note that

$$\sum_{i=1}^{n}(x_i + \alpha y_i)^2 \geq 0, \quad \forall\, \alpha \in \mathbb{R}. \qquad (3.45)$$

Also,

$$\sum_{i=1}^{n}(x_i + \alpha y_i)^2 = \alpha^2\left(\sum_{i=1}^{n} y_i^2\right) + 2\alpha\left(\sum_{i=1}^{n} x_i\, y_i\right) + \left(\sum_{i=1}^{n} x_i^2\right). \tag{3.46}$$

From (3.45) and (3.46), it follows that

$$\alpha^2\left(\sum_{i=1}^{n} y_i^2\right) + 2\alpha\left(\sum_{i=1}^{n} x_i\, y_i\right) + \left(\sum_{i=1}^{n} x_i^2\right) \geq 0, \tag{3.47}$$

for all $\alpha \in \mathbb{R}$.

Note that the left hand side of (3.47) is a quadratic polynomial of α. The inequality (3.47) holds true for any real number α if and only if this polynomial has at most one real double root, i.e., if and only if its discriminant is less than of equal to 0:

$$\left(2\left(\sum_{i=1}^{n} x_i\, y_i\right)\right)^2 - 4\left(\sum_{i=1}^{n} x_i^2\right)\left(\sum_{i=1}^{n} y_i^2\right) \leq 0.$$

This is equivalent to

$$\left(\sum_{i=1}^{n} x_i^2\right)\left(\sum_{i=1}^{n} y_i^2\right) \geq \left(\sum_{i=1}^{n} x_i y_i^2\right)^2,$$

for all $x_i, y_i \in \mathbb{R}$, $i = 1 : n$, and which is the Cauchy–Schwartz inequality. □

Question 19. Compute

$$\int_{-\infty}^{\infty} e^{-x^2}\, dx.$$

Answer: The fastest and most natural way to calculate this integral is to recall that the integral of the probability density function of the standard normal distribution is $f(t) = \frac{1}{\sqrt{2\pi}}e^{-\frac{t^2}{2}}$ over the entire real axis is 1, i.e.,

$$\int_{-\infty}^{\infty} f(t)\,dt = \frac{1}{\sqrt{2\pi}}\int_{-\infty}^{\infty} e^{-\frac{t^2}{2}}\,dt = 1.$$

Thus,

$$\int_{-\infty}^{\infty} e^{-\frac{t^2}{2}}\,dt = \sqrt{2\pi}. \tag{3.48}$$

Using the change of variables $t = x\sqrt{2}$ in (3.48), which corresponds to $dt = dx\sqrt{2}$, we obtain that

$$\begin{aligned} \int_{-\infty}^{\infty} e^{-\frac{(x\sqrt{2})^2}{2}}\,(dx\sqrt{2}) &= \sqrt{2\pi}; \\ \Longleftrightarrow \quad \sqrt{2}\int_{-\infty}^{\infty} e^{-x^2}\,dx &= \sqrt{2\pi}, \end{aligned}$$

and we conclude that

$$\int_{-\infty}^{\infty} e^{-x^2}\,dx = \sqrt{\pi}. \quad \square$$

Question 20. Solve the differential equation

$$y'(x) - y(x) = f(x),$$

where $f(x)$ is a continuous function.

Answer: By multiplying the ODE by e^{-x}, we obtain that

$$\begin{aligned} e^{-x}y'(x) - e^{-x}y(x) &= e^{-x}f(x) \\ \Longleftrightarrow \quad \left(e^{-x}y(x)\right)' &= e^{-x}f(x) \\ \Longleftrightarrow \quad e^{-x}y(x) &= \int e^{-x}f(x)\,dx + C, \end{aligned}$$

where C is an arbitrary constant.

Thus, we conclude that

$$y(x) = Ce^x + e^x \int e^{-x} f(x)\, dx. \quad \square \tag{3.49}$$

Question 21. Solve the differential equation

$$y''(x) - y(x) = g(x),$$

where $g(x)$ is a continuous function.

Answer: The ODE $y''(x) - y(x) = g(x)$ can be written as

$$y''(x) - y'(x) + y'(x) - y(x) = g(x),$$

which is equivalent to

$$w'(x) + w(x) = g(x), \tag{3.50}$$

where

$$w(x) = y'(x) - y(x). \tag{3.51}$$

By multiplying (3.50) by e^x, we obtain that

$$\begin{aligned} e^x w'(x) + e^x w(x) &= e^x g(x) \\ \Longleftrightarrow \quad (e^x w(x))' &= e^x g(x) \\ \Longleftrightarrow \quad e^x w(x) &= \int e^x g(x)\, dx + C_1, \end{aligned}$$

where C_1 is an arbitrary constant. Then,

$$w(x) = C_1 e^{-x} + e^{-x} \int e^x g(x)\, dx. \tag{3.52}$$

Recall from (3.51) that

$$y'(x) - y(x) = w(x). \tag{3.53}$$

According to (3.49), the solution to the ODE $y'(x) - y(x) = f(x)$ is

$$y(x) = Ce^x + e^x \int e^{-x} f(x)\, dx,$$

where C is an arbitrary constant. Thus, using the formula (3.52) for $w(x)$, we find that the solution to the ODE (3.53) is

$$\begin{aligned} y(x) &= Ce^x + e^x \int e^{-x} w(x)\,dx \\ &= Ce^x + C_1 e^x \int e^{-2x}\,dx \\ &\quad + e^x \int e^{-2x} \left(\int e^x g(x) dx \right) dx. \end{aligned} \tag{3.54}$$

Note that

$$\begin{aligned} C_1 e^x \int e^{-2x}\,dx &= C_1 e^x \left(-\frac{1}{2} e^{-2x} + \widetilde{C} \right) \\ &= -\frac{C_1}{2} e^{-x} + C_1 \widetilde{C} e^x, \end{aligned}$$

where $\widetilde{C}$ is an arbitrary constant. By denoting $C_2 = -\frac{C_1}{2}$ and $C_3 = C_1\widetilde{C}$, we conclude that

$$C_1 e^x \int e^{-2x}\,dx = C_2 e^{-x} + C_3 e^x, \tag{3.55}$$

where C_2 and C_3 are arbitrary constants.

From (3.54) and (3.55), we conclude that

$$\begin{aligned} y(x) &= Ce^x + C_2 e^{-x} + C_3 e^x \\ &\quad + e^x \int e^{-2x} \left(\int e^x g(x)\,dx \right) dx \\ &\overset{C_4 = C + C_3}{=} C_4 e^x + C_2 e^{-x} \\ &\quad + e^x \int e^{-2x} \left(\int e^x g(x)\,dx \right) dx, \end{aligned}$$

where C_2 and C_4 are arbitrary constants. □

Question 22. Find the value of

$$\int_0^a \frac{f(x)}{f(x) + f(a-x)}\,dx,$$

assuming that $f(x)$ is a function such that the integral above exists.

Answer: Let

$$I = \int_0^a \frac{f(x)}{f(x)+f(a-x)}\, dx. \tag{3.56}$$

We use the substitution $y = a - x$ for (3.56). Note that $x = 0$ corresponds to $y = a$, $x = a$ corresponds to $y = 0$, and, since $x = a - y$, it follows that $dx = -dy$. Then, from (3.56), we obtain that

$$\begin{aligned} I &= \int_0^a \frac{f(x)}{f(x)+f(a-x)}\, dx \\ &= -\int_a^0 \frac{f(a-y)}{f(a-y)+f(y)}\, dy \\ &= \int_0^a \frac{f(a-y)}{f(y)+f(a-y)}\, dy. \end{aligned} \tag{3.57}$$

By renaming the integrating variable y as x in (3.57), we find that

$$I = \int_0^a \frac{f(a-x)}{f(x)+f(a-x)}\, dx. \tag{3.58}$$

By adding (3.56) and (3.58), we obtain that

$$2I = \int_0^a \frac{f(x)+f(a-x)}{f(x)+f(a-x)}\, dx = \int_0^a 1\, dx = a,$$

and conclude that

$$I = \frac{a}{2}.$$

Thus,

$$\int_0^a \frac{f(x)}{f(x)+f(a-x)}\, dx = \frac{a}{2} \tag{3.59}$$

for any function $f(x)$ such that the integral from (3.59) exists. □

3.2 Covariance and correlation matrices. Linear algebra.

Question 1. Show that any covariance matrix is symmetric positive semidefinite. Show that the same is true for correlation matrices.

Answer: Let Σ_X and Ω_X be the covariance matrix and the correlation matrix of n random variables X_1, X_2, ..., X_n. It is easy to see that Σ_X and Ω_X are symmetric matrices:

$$\begin{aligned}\Sigma_X(j,k) &= \operatorname{cov}(X_j,X_k) = \operatorname{cov}(X_k,X_j)\\ &= \Sigma_X(k,j), \ \forall\ 1 \le j,k \le n;\\ \Omega_X(j,k) &= \operatorname{corr}(X_j,X_k) = \operatorname{corr}(X_k,X_j)\\ &= \Omega_X(k,j), \ \forall\ 1 \le j,k \le n.\end{aligned}$$

Let $c_1, c_2, \ldots, c_n$ be real numbers, and let $C = (c_i)_{i=1:n}$ be a column vector of size n. Recall that[1]

$$\operatorname{var}\left(\sum_{i=1}^n c_i X_i\right) = C^t \Sigma_X C. \qquad (3.60)$$

Since the variance of any random variable is nonnegative, it follows that

$$C^t \Sigma_X C \ \ge\ 0, \ \forall\ C \in \mathbb{R}^n, \qquad (3.61)$$

and we conclude that Σ_X is a symmetric positive semidefinite matrix.

[1]Note that (3.60) is a special case of the following more general result: If $C^{(1)} = (c_i^{(1)})_{i=1:n}$ and $C^{(2)} = (c_i^{(2)})_{i=1:n}$ are two column vectors of size n, then,

$$\operatorname{cov}\left(\sum_{i=1}^n c_i^{(1)} X_i, \sum_{i=1}^n c_i^{(2)} X_i\right) = (C^{(1)})^t \Sigma_X C^{(2)};$$

see Lemma 7.3 from Stefanica [5] for a proof of this result.

For completeness, we include a proof of (3.60) here. Let $Y = \sum_{i=1}^n c_i X_i$. Then,

$$Y - E[Y] = \sum_{i=1}^n c_i (X_i - \mu_i),$$

where $\mu_i = E[X_i]$, for $i = 1 : n$, and therefore

$$\begin{aligned}
& \mathrm{var}\left(\sum_{i=1}^n c_i X_i\right) \\
= \; & \mathrm{var}(Y) = E\left[(Y - E[Y])^2\right] \\
= \; & E\left[\left(\sum_{i=1}^n c_i (X_i - \mu_i)\right)^2\right] \qquad (3.62) \\
= \; & E\left[\sum_{1 \le j,k \le n} c_j c_k (X_j - \mu_j)(X_k - \mu_k)\right] \\
= \; & \sum_{1 \le j,k \le n} c_j c_k E[(X_j - \mu_j)(X_k - \mu_k)] \\
= \; & \sum_{1 \le j,k \le n} c_j c_k \mathrm{cov}(X_j, X_k) \\
= \; & \sum_{1 \le j,k \le n} c_j c_k \Sigma_X(j,k) \\
= \; & C^t \Sigma_X C, \qquad (3.63)
\end{aligned}$$

where, for (3.62) and for (3.63), we used the following facts, respectively:

$$\begin{aligned}
\left(\sum_{i=1}^n z_i\right)^2 &= \sum_{1 \le j,k \le n} z_j z_k, \quad \forall \; z_i \in \mathbb{R}, \; i = 1 : n; \\
x^t A x &= \sum_{1 \le j,k \le n} x_j x_k A(j,k),
\end{aligned}$$

for any $n \times 1$ vector $x = (x_i)_{i=1:n} \in \mathbb{R}^n$, and for any $n \times n$ matrix A.

To show that Ω_X is a symmetric positive semidefinite matrix, recall the following correspondence between covariance matrices and correlation matrices:

$$\Sigma_X = D_{\sigma_X} \Omega_X D_{\sigma_X}, \qquad (3.64)$$

where $D_{\sigma_X} = \operatorname{diag}(\sigma_i)_{i=1:n}$ is the diagonal matrix with entries equal to the standard deviations of the n random variables, i.e., $\sigma_i^2 = \operatorname{var}(X_i)$, for $i = 1 : n$.

Note that $(D_{\sigma_X})^{-1} = \operatorname{diag}\left(\frac{1}{\sigma_i}\right)_{i=1:n}$. Let $v \in \mathbb{R}^n$, and let

$$w = (D_{\sigma_X})^{-1} v.$$

Then,

$$w^t = v^t \left((D_{\sigma_X})^{-1}\right)^t = v^t (D_{\sigma_X})^{-1}, \qquad (3.65)$$

since $(D_{\sigma_X})^{-1}$ is a diagonal matrix and therefore symmetric, i.e., $\left((D_{\sigma_X})^{-1}\right)^t = (D_{\sigma_X})^{-1}$.

From (3.64), (3.65), and (3.61), we find that

$$\begin{aligned} w^t \Sigma_X w &= w^t (D_{\sigma_X} \Omega_X D_{\sigma_X}) w \\ &= v^t (D_{\sigma_X})^{-1} (D_{\sigma_X} \Omega_X D_{\sigma_X}) (D_{\sigma_X})^{-1} v \\ &= v^t \Omega_X v \\ &\geq 0. \end{aligned}$$

Thus, $v^t \Omega_X v \geq 0$ for all $v \in \mathbb{R}^n$, and we conclude that Ω_X is a symmetric positive semidefinite matrix. □

Question 2. Find the correlation matrix of three random variables with covariance matrix

$$\Sigma_X = \begin{pmatrix} 1 & 0.36 & -1.44 \\ 0.36 & 4 & 0.80 \\ -1.44 & 0.80 & 9 \end{pmatrix}. \qquad (3.66)$$

Answer: If Ω_X is the correlation matrix of the three random variables, then

$$\Sigma_X = \begin{pmatrix} \sigma_1 & 0 & 0 \\ 0 & \sigma_2 & 0 \\ 0 & 0 & \sigma_3 \end{pmatrix} \Omega_X \begin{pmatrix} \sigma_1 & 0 & 0 \\ 0 & \sigma_2 & 0 \\ 0 & 0 & \sigma_3 \end{pmatrix},$$

where σ_i, $i = 1:3$, are the standard deviations of the three random variables, and therefore

$$\Omega_X = \begin{pmatrix} \frac{1}{\sigma_1} & 0 & 0 \\ 0 & \frac{1}{\sigma_2} & 0 \\ 0 & 0 & \frac{1}{\sigma_3} \end{pmatrix} \Sigma_X \begin{pmatrix} \frac{1}{\sigma_1} & 0 & 0 \\ 0 & \frac{1}{\sigma_2} & 0 \\ 0 & 0 & \frac{1}{\sigma_3} \end{pmatrix}.$$

Since the standard deviations of the random variables with covariance matrix Σ_X given by (3.66) are

$$\begin{aligned} \sigma_1 &= \sqrt{\Sigma_X(1,1)} = 1; \\ \sigma_2 &= \sqrt{\Sigma_X(2,2)} = 2; \\ \sigma_3 &= \sqrt{\Sigma_X(2,2)} = 3, \end{aligned}$$

we obtain that

$$\begin{aligned} \Omega_X &= \begin{pmatrix} 1 & 0 & 0 \\ 0 & \frac{1}{2} & 0 \\ 0 & 0 & \frac{1}{3} \end{pmatrix} \Sigma_X \begin{pmatrix} 1 & 0 & 0 \\ 0 & \frac{1}{2} & 0 \\ 0 & 0 & \frac{1}{3} \end{pmatrix} \\ &= \begin{pmatrix} 1 & 0.18 & -0.48 \\ 0.18 & 1 & 0.1333 \\ -0.48 & 0.1333 & 1 \end{pmatrix}. \quad \square \end{aligned}$$

Question 3. Assume that all the entries of an $n \times n$ correlation matrix which are not on the main diagonal are equal to ρ. Find upper and lower bounds on the possible values of ρ.

Answer: Recall that a symmetric matrix with diagonal entries equal to 1 is a correlation matrix if and only if the matrix is symmetric positive semidefinite, i.e., if and only if all the eigenvalues of the matrix are nonnegative.

Let

$$\Omega = \begin{pmatrix} 1 & \rho & \cdots & \rho \\ \rho & \ddots & \ddots & \vdots \\ \vdots & \ddots & \ddots & \rho \\ \rho & \cdots & \rho & 1 \end{pmatrix}.$$

We include two ways to compute the eigenvalues of Ω, which are then used to find the necessary and sufficient conditions for the matrix Ω to be a correlation matrix.

Solution 1: Note that

$$\begin{aligned} \Omega &= (1-\rho)I + \begin{pmatrix} \rho & \rho & \cdots & \rho \\ \rho & \ddots & \ddots & \vdots \\ \vdots & \ddots & \ddots & \rho \\ \rho & \cdots & \rho & \rho \end{pmatrix} \\ &= (1-\rho)I + \rho M, \end{aligned}$$

where M is the $n \times n$ matrix with all entries equal to 1, and I is the $n \times n$ identity matrix.

Let λ and $v = (v_i)_{i=1:n}$ be an eigenvalue and a corresponding eigenvector of M, i.e., $Mv = \lambda v$, with $v \neq 0$.

Then, $Mv = \lambda v$ can be written as

$$\left\{ \begin{array}{rcl} v_1 + v_2 + \cdots + v_n &=& \lambda v_1; \\ v_1 + v_2 + \cdots + v_n &=& \lambda v_2; \\ \vdots & & \vdots \\ v_1 + v_2 + \cdots + v_n &=& \lambda v_n, \end{array} \right.$$

and therefore

$$\lambda v_1 = \lambda v_2 = \ldots = \lambda v_n$$

Thus, either $\lambda = 0$, or $v_1 = v_2 = \cdots = v_n$, in which case $nv_1 = \lambda v_1$, and therefore $\lambda = n$, since $v = (v_i)_{i=1:n} \neq 0$.

In other words, the eigenvalues of M are $\lambda = 0$ and $\lambda = n$.

Note that, if $Mv = \lambda v$, then,

$$\begin{aligned} \Omega v &= (1-\rho)v + \rho M v = (1-\rho)v + \rho\lambda v \\ &= (1-\rho+\rho\lambda)v. \end{aligned}$$

Thus, $\mu = 1 - \rho + \rho\lambda$ and v are an eigenvalue and corresponding eigenvector of Ω.

Since the eigenvalues of M are $\lambda = 0$ and $\lambda = n$, it follows that the eigenvalues[2] of Ω are $\mu = 1 - \rho$, corresponding to $\lambda = 0$, and $\mu = (1-\rho) + n\rho = 1 + (n-1)\rho$, corresponding to $\lambda = n$.

Since Ω is a correlation matrix if and only if all its eigenvalues are nonnegative, we conclude that the matrix Ω is a correlation matrix if and only if

$$0 \le 1 + (n-1)\rho \quad \text{and} \quad 0 \le 1 - \rho,$$

which is equivalent to

$$-\frac{1}{n-1} \le \rho \le 1. \tag{3.67}$$

Solution 2: Note that

$$\begin{aligned} \Omega &= (1-\rho)I + \rho \begin{pmatrix} 1 & 1 & \dots & 1 \\ 1 & \ddots & \ddots & \vdots \\ \vdots & \ddots & \ddots & 1 \\ 1 & \dots & 1 & 1 \end{pmatrix} \\ &= (1-\rho)I + \rho \begin{pmatrix} 1 \\ \vdots \\ 1 \end{pmatrix} (1 \ \dots \ 1) \\ &= (1-\rho)I + \rho w w^t \\ &= (1-\rho)I + \rho A, \end{aligned}$$

[2]The eigenvalue $1-\rho$ has multiplicity $n-1$, and the eigenvalue $1+(n-1)\rho$ has multiplicity 1; see Solution 2 of this question.

where I is the $n \times n$ identity matrix and A is the $n \times n$ matrix given by $A = ww^t$, where w is the $n \times 1$ column vector of size n with all entries equal to 1.

Recall that an $n \times n$ matrix of the form uu^t, where $u = (u_i)_{i=1:n}$ is an $n \times 1$ column vector, has an eigenvalue equal to $\sum_{i=1}^n u_i^2$ with multiplicity 1 and another eigenvalue equal to 0 with multiplicity $n-1$.

Then, the eigenvalues of the matrix A are:
$\lambda = \sum_{i=1}^n w_i^2 = \sum_{i=1}^n 1 = n$ with multiplicity 1;
$\lambda = 0$ with multiplicity $n-1$.

Note that, if λ and v are an eigenvalue and a corresponding eigenvector of A, then $Av = \lambda v$, and therefore

$$\begin{aligned} \Omega v &= (1-\rho)v + \rho Av \;=\; (1-\rho)v + \rho\lambda v \\ &= (1-\rho+\rho\lambda)v. \end{aligned}$$

Thus, $1-\rho+\rho\lambda$ and v are an eigenvalue and corresponding eigenvector of Ω. and we obtain that the matrix Ω has the following eigenvalues:

- $(1-\rho) + n\rho = 1 + (n-1)\rho$ with multiplicity 1;
- $1-\rho$ with multiplicity $n-1$.

As before, since Ω is a correlation matrix if and only if all its eigenvalues are nonnegative, we conclude that the matrix Ω is a correlation matrix if and only if

$$0 \leq 1 + (n-1)\rho \quad \text{and} \quad 0 \leq 1 - \rho,$$

which is equivalent to

$$-\frac{1}{n-1} \;\leq\; \rho \;\leq\; 1,$$

which is the same as (3.67). $\square$

Question 4. How many eigenvalues does an $n \times n$ matrix with real entries have? How many eigenvectors?

Answer: Any $n \times n$ matrix with real entries has n eigenvalues, counted with their multiplicities; some of the eigenvalues may be complex numbers. Any $n \times n$ matrix has at most n eigenvectors.

Let A be an $n \times n$ matrix. Let λ be an eigenvalue of A with corresponding eigenvector $v \neq 0$, and let $P_A(x) = \det(xI_n - A)$ be the characteristic polynomial of A, where I_n is the $n \times n$ identity matrix. Note that

$$\begin{aligned} Av = \lambda v, \ v \neq 0 \quad &\Longleftrightarrow \quad (\lambda I_n - A)v = 0, \ v \neq 0 \\ &\Longleftrightarrow \quad \lambda I_n - A \text{ singular matrix} \\ &\Longleftrightarrow \quad \det(\lambda I_n - A) = 0 \\ &\Longleftrightarrow \quad P_A(\lambda) = 0. \end{aligned}$$

In other words, λ is an eigenvalue of A if and only if λ is a root of the corresponding characteristic polynomial $P_A(x)$. Since $P_A(x)$ is a polynomial of degree n, it follows from the Fundamental Theorem of Algebra that $P_A(x)$ has exactly n (complex) roots when counted with their multiplicities. We conclude that any $n \times n$ matrix has n eigenvalues, counted with their multiplicities.

An eigenvalue of multiplicity m has at least one eigenvector and at most m linearly independent corresponding eigenvectors, but it may have less than m linearly independent eigenvectors.[3] Thus, an $n \times n$ matrix has at most n eigenvectors, and at least as many eigenvectors as the number of distinct eigenvalues of the matrix. □

Question 5. Let

$$A = \begin{pmatrix} 2 & -2 \\ -2 & 5 \end{pmatrix}.$$

[3] For example, the matrix $\begin{pmatrix} 2 & 1 & 0 \\ 0 & 2 & 1 \\ 0 & 0 & 2 \end{pmatrix}$ has eigenvalue 2 with multiplicity 3 and only one eigenvector, $\begin{pmatrix} 1 \\ 0 \\ 0 \end{pmatrix}$.

(i) Find a 2×2 matrix M such that $M^2 = A$;

(ii) Find a 2×2 matrix M such that $A = MM^t$.

Answer: (i) Recall that any symmetric matrix has the diagonal form

$$A = Q\Lambda Q^t, \tag{3.68}$$

where Λ is the diagonal matrix whose entries on the main diagonal are the eigenvalues of A and Q is the orthogonal matrix whose columns are the corresponding eigenvectors of A of norm 1, i.e.,

$$\Lambda = \begin{pmatrix} \lambda_1 & 0 \\ 0 & \lambda_2 \end{pmatrix}; \quad Q = \begin{pmatrix} v_1 & v_2 \end{pmatrix}, \tag{3.69}$$

where $Av_1 = \lambda_1 v_1$ and $Av_2 = \lambda_2 v_2$, with $||v_1|| = ||v_2|| = 1$.

If the matrix A has nonnegative eigenvalues, i.e., if $\lambda_1 \geq 0$ and $\lambda_2 \geq 0$, then the matrix

$$M = Q\Lambda^{1/2}Q^t \tag{3.70}$$

with

$$\Lambda^{1/2} = \begin{pmatrix} \sqrt{\lambda_1} & 0 \\ 0 & \sqrt{\lambda_2} \end{pmatrix} \tag{3.71}$$

has the property that $M^2 = A$:

$$\begin{aligned} M^2 &= \left(Q\Lambda^{1/2}Q^t\right)\left(Q\Lambda^{1/2}Q^t\right) \\ &= Q\Lambda^{1/2}(Q^tQ)\Lambda^{1/2}Q^t \\ &= Q\Lambda^{1/2}\Lambda^{1/2}Q^t \\ &= Q\Lambda Q^t \\ &= A, \end{aligned}$$

since Q is an orthogonal matrix and therefore $Q^tQ = I$, and since, from (3.71), it follows that $\Lambda^{1/2}\Lambda^{1/2} = \Lambda$.

We now proceed to compute the eigenvalues and the eigenvectors of the matrix A. The eigenvalues of the matrix A are the roots of the characteristic polynomial $P_A(x)$ of the matrix A given by[4]

$$\begin{aligned} P_A(x) &= \det(xI - A) = \det\begin{pmatrix} x-2 & 2 \\ 2 & x-5 \end{pmatrix} \\ &= (x-2)(x-5) - 4 = x^2 - 7x + 6 \\ &= (x-1)(x-6). \end{aligned}$$

The roots of $P_A(x)$ are 1 and 6, and therefore the eigenvalues of A are $\lambda_1 = 1$ and $\lambda_2 = 6$. The corresponding eigenvectors of norm 1 are

$$v_1 = \begin{pmatrix} \frac{2}{\sqrt{5}} \\ \frac{1}{\sqrt{5}} \end{pmatrix} \quad \text{and} \quad v_2 = \begin{pmatrix} \frac{1}{\sqrt{5}} \\ -\frac{2}{\sqrt{5}} \end{pmatrix}.$$

For example, if $\lambda_2 = 6$, any corresponding eigenvector $v_2 = \begin{pmatrix} a \\ b \end{pmatrix} \neq 0$ is a solution to $Av = 6v$, which can be written as

$$\begin{cases} 2a - 2b = 6a \\ -2a + 5b = 6b \end{cases} \iff \begin{cases} b = -2a \\ b = -2a \end{cases}$$

Thus, any eigenvector corresponding to the eigenvalue $\lambda_2 = 6$ is of the form

$$v_2 = \begin{pmatrix} a \\ -2a \end{pmatrix} = a\begin{pmatrix} 1 \\ -2 \end{pmatrix}.$$

By choosing $a = \frac{1}{\sqrt{5}}$, we obtain that an eigenvector of norm 1 corresponding to the eigenvalue $\lambda_2 = 6$ is

$$v_2 = \begin{pmatrix} \frac{1}{\sqrt{5}} \\ -\frac{2}{\sqrt{5}} \end{pmatrix}.$$

---

[4]The characteristic polynomial of the matrix A can also be obtained as follows:

$$P_A(x) = x^2 - \operatorname{tr}(A)x + \det(A) = x^2 - 7x + 6,$$

where $\operatorname{tr}(A) = 2 + 5 = 7$ and $\det(A) = 2 \cdot 5 - (-2) \cdot (-2) = 6$.

Then, it follows from (3.70) and (3.71) that the matrix M given by

$$\begin{aligned} M &= \begin{pmatrix} \frac{2}{\sqrt{5}} & \frac{1}{\sqrt{5}} \\ \frac{1}{\sqrt{5}} & -\frac{2}{\sqrt{5}} \end{pmatrix} \begin{pmatrix} \sqrt{1} & 0 \\ 0 & \sqrt{6} \end{pmatrix} \begin{pmatrix} \frac{2}{\sqrt{5}} & \frac{1}{\sqrt{5}} \\ \frac{1}{\sqrt{5}} & -\frac{2}{\sqrt{5}} \end{pmatrix} \\ &= \frac{1}{5} \begin{pmatrix} 2 & 1 \\ 1 & -2 \end{pmatrix} \begin{pmatrix} 1 & 0 \\ 0 & \sqrt{6} \end{pmatrix} \begin{pmatrix} 2 & 1 \\ 1 & -2 \end{pmatrix} \\ &= \frac{1}{5} \begin{pmatrix} 4+\sqrt{6} & 2-2\sqrt{6} \\ 2-2\sqrt{6} & 1+4\sqrt{6} \end{pmatrix} \end{aligned}$$

has the property that $M^2 = A$.

(ii) We found that the eigenvalues of A are 1 and 6, i.e., positive. Then, A is a symmetric positive definite matrix, and therefore has a Cholesky decomposition. Recall that, if U is the Cholesky factor of the matrix A, then, by definition, $A = U^tU$. Thus, in order to find a matrix M such that $A = MM^t$, it is enough to compute the Cholesky factor

$$U = \begin{pmatrix} U(1,1) & U(1,2) \\ 0 & U(2,2) \end{pmatrix}$$

of the matrix A and let $M = U^t$.

Note that $A = U^tU$ can be written as

$$\begin{aligned} &\begin{pmatrix} A(1,1) & A(1,2) \\ A(2,1) & A(2,2) \end{pmatrix} \qquad (3.72) \\ &= \begin{pmatrix} U(1,1) & 0 \\ U(1,2) & U(2,2) \end{pmatrix} \begin{pmatrix} U(1,1) & U(1,2) \\ 0 & U(2,2) \end{pmatrix}. \end{aligned}$$

In the first step of the Cholesky decomposition, the first row of U is computed as follows:

$$\begin{aligned} U(1,1) &= \sqrt{A(1,1)} = \sqrt{2}; \\ U(1,2) &= \frac{A(1,2)}{U(1,1)} = \frac{-2}{\sqrt{2}} = -\sqrt{2}; \end{aligned}$$

see also (3.72).

In the next step of the Cholesky decomposition, the entry $A(2,2)$ of A is updated to $A(2,2) - U(1,2)^2 = 3$, and therefore

$$\begin{aligned} U(2,2) &= \sqrt{A(2,2) - (U(1,2))^2} \\ &= \sqrt{5 - (-\sqrt{2})^2} \\ &= \sqrt{3} \end{aligned}$$

Thus, the Cholesky factor of the matrix A is

$$U = \begin{pmatrix} \sqrt{2} & -\sqrt{2} \\ 0 & \sqrt{3} \end{pmatrix},$$

and therefore the matrix $M = U^t$ given by

$$M = \begin{pmatrix} \sqrt{2} & 0 \\ -\sqrt{2} & \sqrt{3} \end{pmatrix},$$

has the property that $A = MM^t$. □

Question 6. The 2×2 matrix A has eigenvalues 2 and -3 with corresponding eigenvectors $\begin{pmatrix} 1 \\ 2 \end{pmatrix}$ and $\begin{pmatrix} -1 \\ 3 \end{pmatrix}$. If $v = \begin{pmatrix} 3 \\ 1 \end{pmatrix}$, find Av.

Answer: Let $\lambda_1 = 2$, $v_1 = \begin{pmatrix} 1 \\ 2 \end{pmatrix}$, and $\lambda_2 = -3$, $v_2 = \begin{pmatrix} -1 \\ 3 \end{pmatrix}$. We first find constants $c_1, c_2 \in \mathbb{R}$ such that $v = c_1 v_1 + c_2 v_2$, i.e., such that

$$\left\{ \begin{array}{rcl} 3 &=& c_1 - c_2 \\ 1 &=& 2c_1 + 3c_2 \end{array} \right.$$

The solution of this linear system is $c_1 = 2$ and $c_2 = -1$. Thus,

$$v = 2v_1 - v_2. \tag{3.73}$$

Since $Av_1 = \lambda_1 v_1 = 2v_1$ and $Av_2 = \lambda_2 v_2 = -3v_2$, we find from (3.73) that

$$\begin{aligned} Av &= 2Av_1 - Av_2 = 2(2v_1) - (-3v_2) \\ &= 4v_1 + 3v_2 \\ &= \begin{pmatrix} 1 \\ 17 \end{pmatrix}. \quad \square \end{aligned}$$

Question 7. Let A and B be square matrices of the same size. Show that the traces of the matrices AB and BA are equal.

Answer: Recall that, for any two square matrices A and B of the same size, the matrices AB and BA have the same characteristic polynomial, i.e.,

$$\begin{aligned} P_{AB}(x) &= \det(xI - AB) = \det(xI - BA) \\ &= P_{BA}(x), \quad \forall\, x \in \mathbb{R}, \end{aligned} \tag{3.74}$$

where I is the identity matrix of the same size as the matrices A and B.

Also, recall the following connection between the characteristic polynomial $P_M(x)$ of an $n \times n$ matrix M and the trace $\operatorname{tr}(M)$ of the matrix:

$$\begin{aligned} & P_M(x) \\ = \; & \det(xI - M) \\ = \; & x^n - \operatorname{tr}(M)x^{n-1} + \cdots + (-1)^n \det(M). \end{aligned} \tag{3.75}$$

Since $P_{AB}(x) = P_{BA}(x)$, see (3.74), we conclude from (3.75) that

$$\operatorname{tr}(AB) = \operatorname{tr}(BA). \tag{3.76}$$

For completeness, we include a proof of (3.74). If the matrix B is nonsingular, then

$$xI - AB = B^{-1}(xI - BA)B,$$

and therefore

$$\begin{aligned}\det(xI - AB) &= \det(B^{-1})\det(xI - BA)\det(B)\\ &= \det(xI - BA), \qquad (3.77)\end{aligned}$$

since

$$\det(B^{-1})\det(B) = \det(B^{-1}B) = \det(I) = 1.$$

If the matrix B is singular, let ϵ be a real number, and note that the matrix $B - \epsilon I$ is singular if and only if ϵ is equal to an eigenvalue of B. Since the $n \times n$ matrix B has at most n eigenvalues, it follows that, except for a finite number of values of ϵ, the matrix $B - \epsilon I$ is nonsingular, in which case we obtain from (3.77) that

$$\det(xI - A(B - \epsilon I)) = \det(xI - (B - \epsilon I)A). \qquad (3.78)$$

Since both sides of (3.78) are polynomials of degree n in ϵ, and therefore continuous functions of ϵ, we can let $\epsilon \to 0$ in (3.78) and obtain that

$$\begin{aligned}&\lim_{\epsilon\to 0}\left(\ \det(xI - A(B - \epsilon I))\ \right)\\ &\quad = \lim_{\epsilon\to 0}\left(\ \det(xI - (B - \epsilon I)A)\ \right)\\ \Longleftrightarrow\quad &\det(xI - AB) = \det(xI - BA). \qquad (3.79)\end{aligned}$$

From (3.77) and (3.79), we conclude that $\det(xI - AB) = \det(xI - BA)$ regardless of whether the matrix B is nonsingular or singular, which concludes the proof of (3.74). □

Question 8. Can you find $n \times n$ matrices A and B such that

$$AB - BA = I_n,$$

where I_n is the identity matrix of size n?

Answer: We give a proof by contradiction. If it were possible to find $n \times n$ matrices A and B such that $AB - BA = I$, then

$$\mathrm{tr}(AB - BA) \;=\; \mathrm{tr}(I_n) \;=\; n. \tag{3.80}$$

However,

$$\mathrm{tr}(AB - BA) \;=\; \mathrm{tr}(AB) - \mathrm{tr}(BA) \;=\; 0, \tag{3.81}$$

since, if A and B are square matrices, then $\mathrm{tr}(AB) = \mathrm{tr}(BA)$; cf. (3.76).

Since (3.80) and (3.81) contradict each other, we conclude that there are no matrices A and B such that $AB - BA = I$. $\square$

Question 9. A probability matrix is a matrix with nonnegative entries such that the sum of the entries in each row of the matrix is equal to 1. Show that the product of two probability matrices is a probability matrix.

Answer: We first establish the following equivalent definition for a probability matrix:

The $n \times n$ matrix M is a probability matrix if and only if all the entries of M are nonnegative and

$$M\mathbf{1} \;=\; \mathbf{1}, \tag{3.82}$$

where $\mathbf{1}$ is the $n \times 1$ column vector with all entries equal to 1.

To see this, let $M = \mathrm{row}\,(r_j)_{j=1:n}$ be the row form of the matrix M, where r_j is an $1 \times n$ row vector, for $j = 1 : n$. The sum of all the entries in the j-th row r_j of M can be written as follows:[5]

$$\sum_{k=1}^{n} r_j(k) \;=\; r_j\mathbf{1}. \tag{3.83}$$

[5]Note that r_j is an $1 \times n$ vector and $\mathbf{1}$ is an $n \times 1$ vector, and therefore the expression $r_j\mathbf{1}$ from (3.83) is consistent.

Thus, the definition of a probability matrix as a matrix with the sum of the entries in each row equal to 1 can be written as

$$\begin{aligned}
\sum_{k=1}^{n} r_j(k) &= 1, \;\; \forall\, j = 1:n \\
\iff \;\; r_j\mathbf{1} &= 1, \;\; \forall\, j = 1:n \\
\iff \;\; (r_j\mathbf{1})_{j=1:n} &= \mathbf{1} \\
\iff \;\; M\mathbf{1} &= \mathbf{1},
\end{aligned}$$

since $M\mathbf{1} = (r_j\mathbf{1})_{j=1:n}$ if $M = \mathrm{row}\,(r_j)_{j=1:n}$ is the row form of M.

In other words, we established that (3.82) is an equivalent condition for M to be a probability matrix.

Let A and B be probability matrices. Then all the entries of A and B are nonnegative, and therefore all the entries of AB are also nonnegative. From (3.82), it follows that

$$A\mathbf{1} = \mathbf{1} \quad \text{and} \quad B\mathbf{1} = \mathbf{1},$$

and therefore

$$(AB)\mathbf{1} \;=\; A(B\mathbf{1}) \;=\; A\mathbf{1} \;=\; \mathbf{1}.$$

Then, from (3.82), we conclude that AB is a probability matrix. □

Question 10. Find all the values of ρ such that

$$\begin{pmatrix} 1 & 0.6 & -0.3 \\ 0.6 & 1 & \rho \\ -0.3 & \rho & 1 \end{pmatrix}$$

is a correlation matrix.

Answer: Recall that a solution to this question based on was Sylvester's criterion was included in Chapter 1. We give two more solutions to this question here, one using the Cholesky decomposition, and another one based on the definition of symmetric positive semidefinite matrices.

A symmetric matrix with diagonal entries equal to 1 is a correlation matrix if and only if the matrix is symmetric positive semidefinite. Thus, we need to find all the values of ρ such that the matrix

$$\Omega = \begin{pmatrix} 1 & 0.6 & -0.3 \\ 0.6 & 1 & \rho \\ -0.3 & \rho & 1 \end{pmatrix} \qquad (3.84)$$

is symmetric positive semidefinite.

Solution 1: To identify the values of ρ such that the matrix Ω is symmetric positive semidefinite, we apply the first step of the Cholesky algorithm to Ω, and obtain the following 2×2 matrix:

$$\begin{aligned} & \begin{pmatrix} 1 & \rho \\ \rho & 1 \end{pmatrix} - \begin{pmatrix} 0.6 \\ -0.3 \end{pmatrix} (0.6 \ -0.3) \\ = & \begin{pmatrix} 1 & \rho \\ \rho & 1 \end{pmatrix} - \begin{pmatrix} 0.36 & -0.18 \\ -0.18 & 0.09 \end{pmatrix} \\ = & \begin{pmatrix} 0.64 & \rho + 0.18 \\ \rho + 0.18 & 0.91 \end{pmatrix}. \end{aligned}$$

Thus, the matrix Ω is symmetric positive semidefinite if and only if the matrix

$$M = \begin{pmatrix} 0.64 & \rho + 0.18 \\ \rho + 0.18 & 0.91 \end{pmatrix}$$

is symmetric positive semidefinite. Since $M(1,1) = 0.64 > 0$, it follows that M is symmetric positive semidefinite if and only if $\det(M) \geq 0$, i.e., if and only if

$$\det(M) = 0.5824 - (\rho + 0.18)^2 \geq 0, \qquad (3.85)$$

which is equivalent to

$$|\rho + 0.18| \leq \sqrt{0.5824} = 0.7632.$$

We conclude that Ω is a symmetric positive semidefinite matrix, and therefore a correlation matrix, if and only if

$$-0.7632 \leq \rho + 0.18 \leq 0.7632,$$

which can be written as

$$-0.9432 \leq \rho \leq 0.5832. \tag{3.86}$$

Note that condition (3.86) is the same as condition (1.8) obtained when solving the same question using Sylvester's Criterion; see Chapter 1.

Solution 2: By definition, the matrix Ω is symmetric positive semidefinite if and only if $x^t \Omega x \geq 0$ for all $x = (x_i)_{i=1:3} \in \mathbb{R}^3$. Note that

$$\begin{aligned}
& x^t \Omega x \\
= \quad & (x_1 \ x_2 \ x_3) \begin{pmatrix} 1 & 0.6 & -0.3 \\ 0.6 & 1 & \rho \\ -0.3 & \rho & 1 \end{pmatrix} \begin{pmatrix} x_1 \\ x_2 \\ x_3 \end{pmatrix} \\
= \quad & x_1^2 + x_2^2 + x_3^2 + 1.2 x_1 x_2 - 0.6 x_1 x_3 + 2\rho x_2 x_3.
\end{aligned}$$

By completing the square, we obtain that

$$\begin{aligned}
& x^t \Omega x \\
= \quad & x_1^2 + 2x_1(0.6x_2 - 0.3x_3) + x_2^2 + x_3^2 + 2\rho x_2 x_3 \\
= \quad & (x_1 + 0.6x_2 - 0.3x_3)^2 \\
& -(0.6x_2 - 0.3x_3)^2 + x_2^2 + x_3^2 + 2\rho x_2 x_3 \\
= \quad & (x_1 + 0.6x_2 - 0.3x_3)^2 \\
& +0.64x_2^2 + 2x_2 x_3(\rho + 0.18) + 0.91x_3^2.
\end{aligned}$$

By completing the square once again, we find that

$$\begin{aligned}
& 0.64x_2^2 + 2x_2x_3(\rho + 0.18) + 0.91x_3^2 \\
= \quad & \left(0.8x_2 + x_3\frac{\rho + 0.18}{0.8}\right)^2 \\
& -x_3^2\frac{(\rho + 0.18)^2}{0.64} + 0.91x_3^2 \\
= \quad & \left(0.8x_2 + x_3\frac{\rho + 0.18}{0.8}\right)^2 \\
& +\frac{x_3^2}{0.64}\left(0.5824 - (\rho + 0.18)^2\right),
\end{aligned}$$

and therefore

$$\begin{aligned}
& x^t\Omega x \\
= \quad & (x_1 + 0.6x_2 - 0.3x_3)^2 \\
& + \left(0.8x_2 + x_3\frac{\rho + 0.18}{0.8}\right)^2 \\
& + \frac{x_3^2}{0.64}\left(0.5824 - (\rho + 0.18)^2\right).
\end{aligned}$$

Thus, $x^t\Omega x \geq 0$ for all $x = (x_i)_{i=1:3} \in \mathbb{R}^3$ if and only if

$$\begin{aligned}
& (x_1 + 0.6x_2 - 0.3x_3)^2 \\
& + \left(0.8x_2 + x_3\frac{\rho + 0.18}{0.8}\right)^2 \\
& + \frac{x_3^2}{0.64}\left(0.5824 - (\rho + 0.18)^2\right) \\
\geq \quad & 0, \quad \forall\, x_1, x_2, x_3 \in \mathbb{R}.
\end{aligned}$$

The last inequality holds if and only if

$$0.5824 - (\rho + 0.18)^2 \;\geq\; 0, \qquad (3.87)$$

which is the same as (3.85).

We conclude that Ω is a correlation matrix if and only if

$$-0.9432 \leq \rho \leq 0.5832. \quad \square$$

Question 11. Let

$$A = \begin{pmatrix} 2 & -1 \\ -1 & 2 \end{pmatrix}.$$

Compute A^{50}.

Answer: Since the matrix A is symmetric, it has the diagonal form

$$A = Q\Lambda Q^t, \tag{3.88}$$

where Λ is a diagonal matrix whose entries are the eigenvalues of the matrix A and Q is an orthogonal matrix whose columns are the corresponding eigenvectors of the matrix A normalized to norm 1. In other words, if λ_1 and λ_2 are the eigenvalues of the 2×2 matrix A and v_1 and v_2 are the corresponding eigenvectors of norm 1, then

$$\Lambda = \begin{pmatrix} \lambda_1 & 0 \\ 0 & \lambda_2 \end{pmatrix}; \quad Q = [v_1 \;\; v_2]. \tag{3.89}$$

Since $Q^tQ = I$, it follows from (3.88) that

$$A^{50} = Q\Lambda^{50}Q^t. \tag{3.90}$$

The eigenvalues of the matrix A are the roots of its characteristic polynomial

$$\begin{aligned} P_A(t) &= t^2 - \text{trace}(A)\,t + \det(A) \\ &= t^2 - 4t + 3 \\ &= (t-1)(t-3). \end{aligned}$$

By solving $P_A(t) = 0$, we obtain that the eigenvalues of the matrix A are $\lambda_1 = 1$ and $\lambda_2 = 3$. The corresponding eigenvectors v_1 and v_2 can be found by solving $Av_1 = \lambda_1 v_1$ and $Av_2 = \lambda_1 v_2$; we obtain that

$$v_1 = C \begin{pmatrix} 1 \\ 1 \end{pmatrix}; \quad v_2 = C \begin{pmatrix} 1 \\ -1 \end{pmatrix}, \tag{3.91}$$

where C is a real constant. Since the eigenvectors v_1 and v_2 from $Q = [v_1 \; v_2]$ must have norm 1, we choose $C = \frac{1}{\sqrt{2}}$ in (3.91) and obtain the following eigenvectors of norm 1:

$$v_1 = \begin{pmatrix} \frac{1}{\sqrt{2}} \\ \frac{1}{\sqrt{2}} \end{pmatrix}; \quad v_2 = \begin{pmatrix} \frac{1}{\sqrt{2}} \\ -\frac{1}{\sqrt{2}} \end{pmatrix}.$$

Thus, from (3.89), we find that

$$\begin{aligned} \Lambda &= \begin{pmatrix} 1 & 0 \\ 0 & 3 \end{pmatrix}; \\ Q &= \begin{pmatrix} \frac{1}{\sqrt{2}} & \frac{1}{\sqrt{2}} \\ \frac{1}{\sqrt{2}} & -\frac{1}{\sqrt{2}} \end{pmatrix} = \frac{1}{\sqrt{2}} \begin{pmatrix} 1 & 1 \\ 1 & -1 \end{pmatrix}. \end{aligned}$$

From (3.90), we conclude that

$$\begin{aligned} A^{50} &= \frac{1}{2} \begin{pmatrix} 1 & 1 \\ 1 & -1 \end{pmatrix} \begin{pmatrix} 1 & 0 \\ 0 & 3^{50} \end{pmatrix} \begin{pmatrix} 1 & 1 \\ 1 & -1 \end{pmatrix} \\ &= \frac{1}{2} \begin{pmatrix} 1 + 3^{50} & 1 - 3^{50} \\ 1 - 3^{50} & 1 + 3^{50} \end{pmatrix}. \quad \square \end{aligned}$$

Question 12. Let A be an $n \times n$ matrix. How do you compute A^{50} in a numerically efficient way minimizing the roundoff errors? What is the operation count to do so?

Answer: Our goal is to compute one specific power of A, that is, A^{50}. To do this, we compute A^2 by squaring the matrix A, A^4 by squaring the matrix A^2, A^8 by squaring the matrix A^4, A^{16} by squaring the matrix A^8, and A^{32} by squaring the matrix A^{16}. Each time we compute the square of an $n \times n$ matrix, we do $2n^3$ operations; for our purpose here, additions, subtractions, multiplications and divisions all count as one operation. In other words, computing the matrices A^2, A^4, A^8, A^{16}, and A^{32} requires

$$5 \times 2n^3 = 10n^3$$

operations.

The matrix A^{50} can be written as the product[6]

$$A^{50} = A^{32} \cdot A^{16} \cdot A^2,$$

involving two $n \times n$ matrix multiplications which require

$$2 \times 2n^3 = 4n^3$$

operations.

Altogether, computing the matrix A^{50} for the $n \times n$ matrix A requires

$$10n^3 + 4n^3 = 14n^3$$

operations. □

Question 13. Let X_1, X_2, and X_3 be three random variables on the same probability space. If the correlation between X_1 and X_3 is 0, what is the possible range for the correlations between X_1 and X_2 and between X_2 and X_3?

Answer: Denote by $\rho_{1,2}$ the correlation between X_1 and X_2 and by $\rho_{2,3}$ the correlation between X_2 and X_3. Since the correlation between X_1 and X_3 is 0, the correlation matrix Ω of the three random variables X_1, X_2, X_3 is

$$\Omega = \begin{pmatrix} 1 & \rho_{1,2} & 0 \\ \rho_{1,2} & 1 & \rho_{2,3} \\ 0 & \rho_{2,3} & 1 \end{pmatrix}.$$

[6]This corresponds to writing the exponent 50 in base 2:

$$50 = \overline{110010} = 2^5 + 2^4 + 2^1 = 32 + 16 + 2.$$

Recall that a matrix is a correlation matrix if and only if the matrix is symmetric positive semidefinite (spsd); see, e.g., Theorem 7.5 from Stefanica [5]. Thus, the possible range for the correlations $\rho_{1,2}$ and $\rho_{2,3}$ is the same as the range for $\rho_{1,2}$ and $\rho_{2,3}$ that makes the matrix Ω symmetric positive semidefinite.

Furthermore, recall from Simpson's Criterion that a matrix is symmetric positive semidefinite if and only if all the principal minors of the matrix are nonnegative.

In other words, our problem is equivalent to finding the ranges of $\rho_{1,2}$ and $\rho_{2,3}$ such that all the principal minors of the matrix Ω are nonnegative.

The matrix Ω has the following seven principal minors: 1, 1, 1,

$$\begin{aligned}
\det\begin{pmatrix} 1 & 0 \\ 0 & 1 \end{pmatrix} &= 1 > 0; \\
\det\begin{pmatrix} 1 & \rho_{1,2} \\ \rho_{1,2} & 1 \end{pmatrix} &= 1 - \rho_{1,2}^2; \\
\det\begin{pmatrix} 1 & \rho_{2,3} \\ \rho_{2,3} & 1 \end{pmatrix} &= 1 - \rho_{2,3}^2; \\
\det(\Omega) &= 1 - \rho_{1,2}^2 - \rho_{2,3}^2.
\end{aligned}$$

Thus, the principal minors of Ω are nonnegative if and only if

$$\rho_{1,2}^2 \leq 1; \quad \rho_{2,3}^2 \leq 1; \quad \rho_{1,2}^2 + \rho_{2,3}^2 \leq 1.$$

This corresponds to

$$\begin{aligned}
-1 \;\leq\; &\rho_{2,3} \;\leq\; 1; \\
-\sqrt{1-\rho_{2,3}^2} \;\leq\; &\rho_{1,2} \;\leq\; \sqrt{1-\rho_{2,3}^2}.
\end{aligned}$$ □

Question 14. Let A and B be square matrices of the same size such that $AB = BA$ and $B^3 = 0$. If the matrix $I + AB$ is nonsingular, find the inverse matrix of $I + AB$.

Answer: Since $(1+x)(1-x+x^2) = 1+x^3$, it follows that

$$(I + AB)\,(I - AB + (AB)^2) \;=\; I + (AB)^3. \qquad (3.92)$$

Using the facts that $AB = BA$ and $B^3 = 0$, we find that

$$(AB)^3 \;=\; AB \cdot AB \cdot AB \;=\; A^3 \cdot B^3 \;=\; 0. \qquad (3.93)$$

From (3.92) and (3.93), we obtain that

$$(I + AB)\,(I - AB + (AB)^2) \;=\; I.$$

Thus,

$$\begin{aligned}(I + AB)^{-1} &= I - AB + (AB)^2 \\ &= I - AB + AB \cdot AB \\ &= I - AB + A^2B^2,\end{aligned}$$

where, for the last equality, we used the fact that $AB = BA$.

We conclude that the inverse matrix of $I + AB$ is

$$I - AB + A^2B^2. \quad \square$$

3.3 Financial instruments: options, bonds, swaps, forwards, futures.

Question 1. The prices of three put options with strikes 40, 50, and 70, but otherwise identical, are \$10, \$20, and \$30, respectively. Is there an arbitrage opportunity present? If yes, how can you make a riskless profit?

Answer: If an arbitrage exists, it will be due to the fact that the convexity of put option values with respect to the strike price is violated.

In the plane (K, y), the line passing through the points $(K = 40, P(40) = 10)$ and $(K = 70, P(70) = 30)$ is given by

$$y = \frac{70 - K}{30} \cdot 10 + \frac{K - 40}{30} \cdot 30. \qquad (3.94)$$

The point on this line corresponding to strike 50 is obtained by substituting $K = 50$ in (3.94), and has y–coordinate equal to

$$\frac{2}{3} \cdot 10 + \frac{1}{3} \cdot 30 = \frac{50}{3}.$$

Since $P(K)$ is a strictly convex function of K, a no–arbitrage value of the put option with strike 50 should be below the line passing through the price points of the options with strikes 40 and 70. However, $P(50) = 20 > \frac{50}{3}$. Thus, the put option with strike 50 is overpriced, and an arbitrage exists.

Using a "buy low, sell high" strategy, we can take advantage of this arbitrage opportunity as follows: buy 2 put options with strike 40, buy 1 put option with strike 70, and sell 3 put options with strike 50. There is a \$10 positive cash flow when setting up this portfolio, since

$$3 \cdot \$20 - 2 \cdot \$10 - \$30 = \$10.$$

The value $V(T)$ of the portfolio at the maturity T of the options is

$$\begin{aligned} V(T) &= 2\max(40 - S(T), 0) \\ &\quad + \max(70 - S(T), 0) \\ &\quad - 3\max(50 - S(T), 0). \end{aligned}$$

Note that $V(T)$ is nonnegative for any value $S(T)$ of the underlying asset at T:

If $70 \leq S(T)$, then all options expire out of the money and

$$V(T) = 0.$$

If $50 \leq S(T) < 70$, then

$$V(T) = 70 - S(T) \geq 0.$$

If $40 \leq S(T) < 50$, then

$$\begin{aligned} V(T) &= (70 - S(T)) - 3(50 - S(T)) \\ &= 2S(T) - 80 \\ &\geq 0. \end{aligned}$$

If $S(T) < 40$, then

$$V(T) = 2S(T) - 80 + 2(40 - S(T)) = 0.$$

In other words, we set up a portfolio with positive cash flow at inception which does not lose money regardless of the value of the underlying asset at time T. The risk–free profit is equal to the future value at time T of the \$10 cash flow from setting up the portfolio. □

Question 2. The price of a stock is \$50. In three months, it will either be \$47 or \$52, with 50% probability. How much would you pay for an at the money put? Assume for simplicity that the stock pays no dividends and that interest rates are zero.

Answer: Recall first that real world probabilities do not play any role in valuing an option in a (one period) binomial tree model. Thus, the 50% probability stated in the question is only meant to throw you off–course.

Solution 1: The value of the option is the discounted expected value of the payoff of the option in the risk–neutral probability measure. Since interest rates are zero, this can be written as

$$P(0) \;=\; p_{RN,up}P_{up} \;+\; p_{RN,down}P_{down}. \qquad (3.95)$$

The up and down factors are $u = \frac{52}{50} = 1.04$ and $d = \frac{47}{50} = 0.94$, respectively. The risk–neutral probabilities of going up and down are

$$p_{RN,up} = \frac{1-d}{u-d} = 0.6; \quad p_{RN,down} = \frac{u-1}{u-d} = 0.4.$$

The ATM put pays \$3 if the stock price goes down to \$47, i.e., $P_{down} = 3$, and expires worthless if the stock price goes up to \$52, i.e., $P_{up} = 0$. From (3.95), we find that the value of the ATM put is

$$P(0) \;=\; 0.6 \cdot 0 + 0.4 \cdot 3 \;=\; 1.2. \qquad (3.96)$$

In other words, you should pay at most \$1.20 for an at the money put.

Solution 2: An insightful solution can be given by setting up a hedged portfolio. The Delta of the put option is

$$\Delta_P \;=\; \frac{P_{up} - P_{down}}{S_{up} - S_{down}} \;=\; \frac{0-3}{52-47} \;=\; -0.6.$$

A portfolio which is long one ATM put and short Δ_P shares will be long the put and long 0.6 shares, and will have the same value at maturity regardless of whether the stock price goes down to \$47 or up to \$52:

$$\begin{aligned} \Pi(T) \;&=\; P(T) + 0.6S(T) \\ &=\; \left\{ \begin{array}{lcl} 0 + 0.6 \cdot 52 &=& 31.20, \text{ if } S(T) = 52; \\ 3 + 0.6 \cdot 47 &=& 31.20, \text{ if } S(T) = 47. \end{array} \right. \end{aligned}$$

For no–arbitrage, the value of the portfolio at inception must be the discounted value of its payoff. Since the interest rates are zero, we obtain that $\Pi(0) = P(0)+0.6S(0) = 31.20$, and therefore

$$P(0) \;=\; 31.20 - 0.6 \cdot 50 \;=\; 1.20,$$

which is the same value of the put, \$1.20, obtained above; see (3.96). □

Question 3. A stock worth \$50 today will be worth either \$60 or \$40 in three months, with equal probability. The value of a three months at the money put on this stock is \$4. Does the value of the three months ATM put increase or decrease, and by how much, if the probability of the stock going up to \$60 were 75% and the probability of the stock going down to \$40 were 25%?

Answer: In a one period binomial tree model, the actual probabilities of the asset going up or down do not play any role in the valuation of a plain vanilla option. Thus, the value of the three months at the money put would be the same, \$4, even if the probability of the stock going up to \$60 were 75%. □

Question 4. What is risk–neutral pricing?

Answer: Risk–neutral pricing, or valuation, refers to valuing derivative securities as discounted expected values of their payoffs at maturity, under the assumption that the underlying asset has lognormal distribution with a drift equal to the risk-free rate.

More precisely, if the price of the underlying asset has a lognormal distribution with volatility σ and pays dividends continuously at the rate q, then the value of a derivative security on this asset with payoff $V(T)$ at maturity T given by risk–neutral valuation is

$$V(0) \;=\; e^{-rT}\; E_{RN}[V(S(T))],$$

where r is the risk–free rate assumed to be constant, and the expected value is computed with respect to the lognormal random variable $S(T)$ given by

$$S(T) = S(0)\, e^{\left(r-q-\frac{\sigma^2}{2}\right)T + \sigma\sqrt{T}Z}.$$

Risk–neutral valuation can be used for derivative securities which can be perfectly hedged dynamically using cash and the underlying asset. Plain vanilla European options, as well as European options with other payoffs at maturity (such as asset–or–nothing and cash–or–nothing options) can be priced using risk–neutrality. Risk–neutral valuation cannot be used for path dependent options such as American options, barrier options, and Asian options. □

Question 5. Describe briefly how you arrive at the Black–Scholes formula.

Answer: The Black–Scholes formulas give the values of plain vanilla European put and call options on an underlying asset with lognormal distribution. Several methods for deriving the Black-Scholes formulas are:

• Risk neutral pricing: the expected value of the payoff of the option at maturity computed under the assumption that the price of the underlying asset has a lognormal distribution with drift equal to the risk free rate gives the Black–Scholes value of the option.

• Black–Scholes PDE solution: the Black–Scholes value of the option satisfies the Black–Scholes PDE with boundary conditions given by the payoff of the option at maturity. The Black–Scholes PDE is transformed into the heat PDE using a lognormal change of variables, and the closed form solution of the heat PDE is then used to derive the closed form solution of the Black–Scholes PDE, which is the Black–Scholes value of the option.

• Binomial tree model pricing: the evolution of the underlying asset is modeled using a binomial tree calibrated to converge in the limit to a lognormal distribution with drift equal to the risk free rate. For every tree, an approximate option value is obtained from the binomial tree model. The limit of these binomial tree option values as the number of time intervals in the tree goes to infinity is the Black–Scholes value of the option.

Note that twelve different ways to derive the Black–Scholes formula can be found in Wilmott [6]. □

Question 6. How much should a three months at the money put on an asset with spot price \$40 and volatility 30% be worth? Assume, for simplicity, that interest rates are zero and that the asset does not pay dividends.

Answer: The following approximation for the value of an at the money put option on a non dividend paying underlying asset and assuming zero risk–free interest rates is easy to estimate and very accurate if the total variance is small (e.g., if $\sigma^2 T \leq 0.25$):

$$P_{ATM} \approx 0.4\sigma S_0 \sqrt{T}; \tag{3.97}$$

see Stefanica [4] for a derivation of formula (3.97).

For $S_0 = 40$, $\sigma = 0.3$, and $T = \frac{1}{4}$, we obtain that the value of the at the money put is approximately 2.40.

For comparison purposes, note that the value of the put option computed using the Black–Scholes formula would be 2.3914; the approximate formula (3.97) is very accurate in this case. □

Question 7. If the price of a stock doubles in one day, by how much will the value of a call option on this stock change?

Answer: The value of a deep in the money call on a non dividend paying asset can be approximated, e.g., by using the Put-Call parity, as $C \approx S - Ke^{-rT}$, where K and T are the strike and the maturity of the option, and r is the constant risk free rate. Thus, if the spot price S doubles, the call option will be even deeper in the money, and therefore its value will be approximately $2S - Ke^{-rT}$. In other words, the value of deep in the money calls roughly doubles if the spot price doubles.

If the option is around at the money, the percentage change generated by the doubling of the stock price is about one order of magnitude larger since the option will become deep in the money.

If the option is deep out of the money, then it trades for fraction of cents. The doubling of the spot price would result in changing of the price of the option by several orders of magnitude.

As a numerical example, consider a six months call option with strike 20 on a non dividend paying underlying asset with volatility 25%. Assume that the risk free rate is constant at 5%. The Black–Scholes values of the call option corresponding to several spot prices of the underlying asset can be found below:

Spot Price	Option Price
10	0.000045
20	1.65
40	20.49
80	60.49
400	380.49
800	780.49

If the call option is deep out of the money and the spot price doubles from \$10 to \$20, the value of the call increases from \$0.000045 to \$1.65, i.e., by more than four orders of magnitude.

If the call option is at the money and the spot price doubles from \$20 to \$40, the value of the call increases from \$1.65 to \$20.49, i.e., more than tenfold.

If the call option is deep in the money, and the spot price doubles from \$40 to \$80, the value of the call increases from \$20.49 to \$60.49, i.e., by a factor of 2.95; if the call option is even deeper in the money[7] and the spot price doubles from \$400 to \$800, the value of the call increases from \$380.49 to \$780.49, i.e., the call approximately doubles in value.

Moreover, if the call option is deep in the money, its value is very close to $S - Ke^{-rT}$, i.e., the value of the spot price of the underlying asset minus the present value of the strike. For all the spot prices greater than \$40, the estimate $C \approx S - Ke^{-rT}$ is very accurate. Thus, if the call option is at the money and the spot price doubles, the value of the call option increases by the same amount as the increase in the spot price. For example, if the spot price doubles from \$40 to \$80, the value of the call increases by \$40, from \$20.49 to \$60.49, which is exactly the increase in the spot price. □

Question 8. What are the smallest and largest values that Delta can take?

Answer: Assume, for simplicity, that the underlying asset does not pay dividends.

The Delta (Δ) of a long position in a plain vanilla call option is between 0 and 1 (and therefore the Delta of a short plain vanilla call position is between -1 and 0). The Delta of a long call position increases with the spot price of the underlying asset, and goes from 0, when the asset is worthless (i.e., when the call option is deep out of the money), to 1, when the spot price of the asset is very large (i.e., when the call option is deep in the money).

[7] Of course, call options with strike ten times smaller than the spot price of the underlying asset never occur in practice; this part of the example is for illustration purposes.

The Delta of a long position in a plain vanilla put option is between -1 and 0; the Delta of a short position in a plain vanilla put option is between 0 and 1. The Delta of a long put position also increases with the spot price of the underlying asset, and goes from -1, when the asset is worthless (i.e., when the put option is deep in the money), to 0, when the spot price of the asset is very large (i.e., when the put option is deep out of the money).

Note that, from the Put-Call parity, i.e., $C - P = S - Ke^{-rT}$, it follows that $\Delta(P) = \Delta(C) - 1$, which is consistent with the bounds above. □

Question 9. What is the Delta of an at–the–money call? What is the Delta of an at–the–money put?

Answer: The Delta of an at–the–money call is approximately 0.5; the Delta of an at–the–money put is approximately -0.5.

Assume, for simplicity, a Black–Scholes framework with zero risk–free rates and an underlying asset paying no dividends. Then, $\Delta(C_{BS}) = N(d_1)$, where $N(x)$ denotes the cumulative distribution of the standard normal variable and

$$d_1 = \frac{\ln\left(\frac{S}{K}\right) + \left(r - q + \frac{\sigma^2}{2}\right)T}{\sigma\sqrt{T}} = \frac{\sigma\sqrt{T}}{2} \tag{3.98}$$

for $K = S$ and for $r = q = 0$.

The linear Taylor approximation of $N(x)$ around 0 is

$$N(x) \approx 0.5 + \frac{x}{\sqrt{2\pi}} \approx 0.5 + 0.4x, \tag{3.99}$$

since $\frac{1}{\sqrt{2\pi}} = 0.3989 \approx 0.4$. Thus,

$$\Delta(C_{BS}) = N(d_1) = N\left(\frac{\sigma\sqrt{T}}{2}\right) \approx 0.5 + 0.2\sigma\sqrt{T}.$$

This is roughly estimated as $\Delta(C_{BS}) \approx 0.5$, since $0.2\sigma\sqrt{T}$ is small for most options (e.g., $0.2\sigma\sqrt{T} \leq 0.1$ for volatility less than 50% and for maturity less than one year).

For put options, $\Delta(P_{BS}) = -N(-d_1)$, where $d_1 = \frac{\sigma\sqrt{T}}{2}$; cf. (3.98). From (3.99), it follows that

$$\Delta(P_{BS}) = -N\left(-\frac{\sigma\sqrt{T}}{2}\right) \approx -0.5 + 0.2\sigma\sqrt{T},$$

which, using the same rationale outlined above, is often stated as $\Delta(P_{BS}) \approx -0.5$. □

Question 10. What is the Put–Call parity? How do you prove it?

Answer: The Put–Call parity is a model independent no–arbitrage relationship between the prices of European call and put options with the same strike and maturity.

In a nutshell, the Put–Call parity states that being long a call option and short a put option on the same underlying asset, and with the same strike and maturity, is the same as being long a forward contract on the asset, with the same maturity as the maturity of the options, and with delivery price equal to the strike of the options. Equivalently, being long a call and short a put is the same as being long one unit of the underlying asset (for non dividend paying assets) and short the present value of the strike of the options.

More precisely, if $C(t)$ and $P(t)$ are the values at time t of a European call and put option with maturity T and strike K, on the same non dividend paying asset with spot price $S(t)$, then the Put–Call parity states that

$$C(t) - P(t) = S(t) - Ke^{-r(T-t)}, \tag{3.100}$$

where r denotes the risk–free rate, assumed to be constant.

If the underlying asset pays dividends continuously at the rate q, the Put–Call parity has the form

$$C(t) \; - \; P(t) \; = \; S(t)e^{-q(T-t)} \; - \; Ke^{-r(T-t)}. \quad (3.101)$$

For simplicity, we restrict our attention to non dividend paying underlying assets and prove formula (3.100).

Consider a portfolio made of the following assets:

- long 1 put option;
- long 1 unit of the underlying asset;
- short 1 call option.

The value of the portfolio at time t is

$$V_{portfolio}(t) \; = \; P(t) + S(t) - C(t). \quad (3.102)$$

The values of the call and put option at maturity are

$$\begin{aligned} C(T) \; &= \; \max(S(T) - K, 0); \\ P(T) \; &= \; \max(K - S(T), 0). \end{aligned}$$

Then, regardless of whether $S(T) < K$ or $S(T) \geq K$, the value of the portfolio at time T will be equal to K:

If $S(T) < K$, then $P(T) = K - S(T)$ and $C(T) = 0$, and therefore

$$\begin{aligned} V_{portfolio}(T) \; &= \; P(T) \; + \; S(T) \; - \; C(T) \\ &= \; (K - S(T)) \; + \; S(T) \; - \; 0 \\ &= \; K. \end{aligned} \quad (3.103)$$

If $S(T) \geq K$, then $P(T) = 0$ and $C(T) = S(T) - K$, and therefore

$$\begin{aligned} V_{portfolio}(T) \; &= \; P(T) \; + \; S(T) \; - \; C(T) \\ &= \; 0 + S(T) - (S(T) - K) \\ &= \; K. \end{aligned} \quad (3.104)$$

From (3.103) and (3.104), we find that

$$V_{portfolio}(T) \; = \; P(T) + S(T) - C(T) \; = \; K, \quad (3.105)$$

regardless of the value $S(T)$ of the underlying asset at the maturity of the option.

Then, for no–arbitrage, the value of the portfolio at time t must be equal to the present value of K at time t, i.e.,

$$V_{portfolio}(t) = Ke^{-r(T-t)}. \tag{3.106}$$

From (3.102) and (3.106), we obtain that

$$P(t) + S(t) - C(t) = Ke^{-r(T-t)}.$$

This can be written as

$$C(t) - P(t) = S(t) - Ke^{-r(T-t)},$$

which is the Put–Call parity formula (3.100).

The Put–Call parity formula (3.101) corresponding to an underlying asset paying dividends continuously at the rate q can be obtained similarly using a portfolio with a long put position, a short call position, and a long position in $e^{-q(T-t)}$ units of the underlying asset. All dividends payed by the long asset position between time t and time T are used to purchase additional fractions of the asset. Doing so continuously results in an asset position at time T equal to long one unit of the asset. Thus, the value of the portfolio at time T will be equal to K regardless of the value of $S(T)$, and the Put–Call parity formula (3.101) for underlying assets paying dividends continuously can be obtained as before. □

Question 11. Show that the time value of a European call option is highest at the money.

Answer: The time value of a call option is the difference between the value $C(S)$ of the option and the intrinsic value $\max(S-K,0)$ of the call option. In other words, the time value of the option is

$$C(S) - \max(S-K,0).$$

We want to establish that the time value of the option is highest at the money, i.e., when $S = K$. To do so, we show that the function

$$f(S) = C(S) - \max(S - K, 0)$$

attains its maximum for $S = K$.

Note that[8]

$$f(S) = \begin{cases} C(S), & \text{if } S \leq K; \\ C(S) - S + K, & \text{if } S > K. \end{cases}$$

For $S \leq K$, the function $f(S)$ is the value of a call with strike K, and therefore is increasing.

For $S > K$, we find that

$$f'(S) = \Delta(C) - 1 < 0,$$

since the Delta of a call option is always less than 1,[9] and therefore the function $f(S)$ is decreasing.

We conclude that $f(S)$ has a global maximum point at $S = K$, which is what we wanted to show.

Note that a similar reasoning shows that the time value of a put option, given by

$$P(S) - \max(K - S, 0),$$

is largest at the money, i.e., for $S = K$. If $g(S) = P(S) - \max(K - S, 0)$, then

$$g(S) = \begin{cases} P(S) - K + S, & \text{if } S \leq K; \\ P(S), & \text{if } S > K. \end{cases}$$

[8]The function $f(S)$ is a continuous function, but it is not differentiable at $S = K$.

[9]The Delta of a long call position is an increasing function going from 0 when the spot price of the underlying asset is 0 and the call option is deep out of the money, to 1 when the spot price of the underlying asset goes to ∞ and the call option is deep in the money. For example, in the Black–Scholes model,

$$\Delta(C_{BS}) = e^{-qT} N(d_1) < N(d_1) < 1.$$

For $S \leq K$,

$$g'(S) = \Delta(P) + 1 > 0,$$

and therefore the function $g(S)$ is increasing, since the Delta of a put option is between -1 (when the put is deep in the money) and 0 (when the put is deep out of the money), and therefore is always greater than -1.

For $S > K$, the function $g(S)$ is the value of a put with strike K, and therefore is decreasing.

Thus, $g(S)$ has a global maximum point at $S = K$, and therefore the time value of the put option is largest at the money. □

Question 12. What is implied volatility? What is a volatility smile? How about a volatility skew?

Answer: By definition, implied volatility is the unique value of the volatility parameter σ from the lognormal model for the evolution of the price of an underlying that makes the Black–Scholes value of an option equal to the market price of the option. Implied volatility exists and is unique[10] for any arbitrage–free market value of the option.

On the same asset, prices of options with multiple strikes and maturities are quoted, and implied volatilities can be computed for each of these options. If the price of the asset had a lognormal distribution as assumed in the Black–Scholes model, then the resulting plots of implied volatility vs strike for the same maturity should be flat. In practice, they are not flat, and are often shaped as "smiles" or "skews".

An implied volatility smile occurs when the implied volatilities of deep in the money options and of deep out of the money options are higher than the implied volatilities

[10]The uniqueness of the implied volatility comes from the fact that the Black–Scholes value of the option is a strictly increasing function of the volatility parameter (or, equivalently, the vega of the option is strictly positive).

of options close to at the money. Volatility smiles are typical for currency options.

An implied volatility skew occurs when the implied volatilities of options with large strikes are lower than the implied volatilities of at the money options (reverse skew), or when the implied volatilities of options with small strikes are lower than the implied volatilities of at the money options (forward skew). Reverse skews are typical for long dated equity and index options. Forward skews are typical for commodities options. □

Question 13. What is the Gamma of an option? Why is it preferable to have small Gamma? Why is the Gamma of plain vanilla options positive?

Answer: The Gamma (Γ) of an option measures the sensitivity of the Delta of the option with respect to the price of the underlying asset, i.e.,

$$\Gamma(V) = \frac{\partial \Delta}{\partial S} = \frac{\partial^2 V}{\partial S^2},$$

where V denotes the value of the option.

It is often important to immunize a portfolio with respect to changes in the price of the underlying asset, i.e., to make the portfolio Delta–neutral. A portfolio with small Gamma would need to be rebalanced less often in order to be kept Delta–neutral, since the change in the Delta of the portfolio is proportional to Gamma for small changes in the value of the underlying asset. Thus, a Delta–neutral and Gamma–neutral portfolio is well hedged against small changes in the price of the underlying asset (although not against jumps in the price of the underlying asset).

The Delta of plain vanilla options (calls or puts) increases as the spot price of the underlying asset increases. Thus, Gamma is positive, since Gamma is the rate of change of Delta. Moreover, Gamma is asymptotically going to 0 for deep out of the money options and for deep in the money options, and the highest value of Gamma corresponds to options with strike close to the spot price (at the money options). □

Question 14. When are a European call and a European put worth the same? (The options are written on the same asset and have the same strike and maturity.) What is the intuition behind this result?

Answer: Recall from the Put–Call parity that

$$C - P = Se^{-qT} - Ke^{-rT}, \tag{3.107}$$

where C and P are the values of a call and put option, respectively, with strike K and maturity T on an underlying asset with spot price S and paying dividends continuously at rate q. If $C = P$, it follows from (3.107) that $K = Se^{(r-q)T}$. Since the forward price of the underlying asset is $F = Se^{(r-q)T}$, we conclude that a call and a put are worth the same if their strike is equal to the forward price of the asset; these options are called at–the–money–forward options.

Note that this result is independent of any assumption on the evolution of the price of the underlying asset, since Put–Call parity is model independent.

The fact that an at–the–money–forward call and an at–the–money–forward put are worth the same may seem counterintuitive at first glance: call options have unlimited upside since their payoff at maturity,

$$C(T) = \max(S(T) - K, 0),$$

can be infinitely large, while put options have limited upside since their payoff at maturity,

$$P(T) = \max(K - S(T), 0),$$

is bounded by the strike price, i.e., $P(T) \leq K$.

However, the value of an option is equal to the risk–neutral expected value of the option payoff at maturity. In every model for the evolution of the price of the underlying asset (including in the geometric Brownian motion model underlying the Black–Scholes framework), the probability density of the underlying asset at maturity decreases exponentially for large values of the spot price. This renders the expected value of large payoffs negligible, and makes it possible for at–the–money–forward put options to be worth the same as at–the–money–forward call options. □

Question 15. What is the two year volatility of an asset with 30% six months volatility?

Answer: Asset volatility scales with the square root of time: if σ denotes the annualized volatility of an asset, the volatility $\sigma(t)$ of the asset over a time horizon t is given by $\sigma(t) = \sigma\sqrt{t}$. Then,

$$\frac{\sigma(t_2)}{\sigma(t_1)} = \sqrt{\frac{t_2}{t_1}}.$$

For $t_2 = 2$, $t_1 = 0.5$, and $\sigma(t_1) = \sigma(0.5) = 0.3$, we find that $\sigma(t_2) = \sigma(2) = 0.6$, i.e., the two year volatility of the asset is 60%. □

Question 16. How do you value an interest rate swap?

Answer: For valuation purposes only, add payments at maturity equal to the notional of the swap both for the fixed leg and for the floating leg of the swap. Then, the value of the swap for the party receiving fixed payments is $V_{swap} = V_{fix} - V_{float}$, where V_{fix} is the value of a coupon bond with coupon rate equal to the fixed rate of the swap, and V_{float} is the value of an instrument making the floating payments of the swap, plus a payment equal to the notional at maturity.

The value V_{fix} of the fixed rate coupon bond is the sum of the present values of its future payments discounted using risk–free zero rates (most frequently, LIBOR rates).

To compute V_{float}, note that, right after a payment is made, the remaining payments of the floating leg (including the notional at maturity) are equivalent to rolling over the notional at the prevailing zero rates until maturity. Thus, the value of all the floating payments on a payment date for the swap is equal to the notional. We conclude that V_{float} is the present value of the next floating payment (which was determined at the prior swap payment date) plus the notional.

To illustrate swap valuation with an example, consider a 19–month semiannual swap on a \$10 million notional with 3% fixed rate and paying semiannually compounded LIBOR. The next floating payment that will be made in one month is \$125,000 (and was determined five months ago, at the previous cash flow date).

The cash flow dates of the swap are 1 month, 7 months, 13 months, and 19 months. The value of the swap for the party receiving fixed payments is $V_{swap} = V_{fix} - V_{float}$. The value of the 3% semiannual coupon bond corresponding to the fixed leg of the swap is

$$\begin{aligned} V_{fix} &= 150,000 \cdot \text{disc}\left(\frac{1}{12}\right) \\ &\quad + 150,000 \cdot \text{disc}\left(\frac{7}{12}\right) \\ &\quad + 150,000 \cdot \text{disc}\left(\frac{13}{12}\right) \\ &\quad + 10,150,000 \cdot \text{disc}\left(\frac{19}{12}\right), \qquad (3.108) \end{aligned}$$

where $\text{disc}(t)$ denotes the discount factor corresponding to time t. For example, if the discount factors are given in terms of the semiannually compounded LIBOR rate LIBOR(t), then

$$\mathrm{disc}(t) = \left(1 + \frac{\mathrm{LIBOR}(t)}{2}\right)^{-2t}.$$

On the next cash flow date, i.e., in one month, the floating leg of the swap is equal to the value of the next floating payment, \$125,000, plus the \$10 million notional, i.e., \$10,125,000. Thus, the value of the floating leg of the swap today is the present value of \$10,125,000 in one month, i.e.,

$$V_{float} = 10,125,000 \cdot \mathrm{disc}\left(\frac{1}{12}\right). \tag{3.109}$$

The value of the swap for the party receiving fixed payments is $V_{swap} = V_{fix} - V_{float}$, where V_{fix} and V_{float} are given by (3.108) and (3.109), respectively. □

Question 17. By how much will the price of a ten year zero coupon bond change if the yield increases by ten basis points?

Answer: The first order approximation of the change ΔB in the bond price in terms of the change Δy in the bond yield is

$$\Delta B \approx \frac{\partial B}{\partial y}\Delta y = -DB\Delta y, \tag{3.110}$$

where $D = -\frac{1}{B}\frac{\partial B}{\partial y}$ is the duration of the bond. Thus,

$$\frac{\Delta B}{B} \approx -D\Delta y. \tag{3.111}$$

Note that $D = 10$, since the duration of a zero coupon bond is equal to the maturity of the bond. Moreover, the change in the bond yield is $\Delta y = 0.001$, since $1\% = 100$ basis points (bps), and therefore 10 bps $= 0.1\% = 0.001$.

From (3.111), we find that the percentage change in the value of the bond can be estimated as follows:

$$\frac{\Delta B}{B} \approx -10 \times 0.001 = -0.01.$$

In other words, the price of the ten year zero-coupon bond decreases by 1% if the yield increases by ten basis points. □

Question 18. A five year bond with 3.5 years duration is worth 102. What is the value of the bond if the yield decreases by fifty basis points?

Answer: The value of the bond will increase, since the yield of the bond decreases. More precisely, recall from (3.110) that

$$\Delta B \approx -DB\Delta y. \tag{3.112}$$

For $B = 102$, $D = 3.5$ and $\Delta y = -0.005$ ($1\% = 100$ bps, and therefore 50 bps $= 0.5\% = 0.005$), we find from (3.112) that $\Delta B \approx 1.785$. Thus, the new value of the bond is $B + \Delta B \approx 103.785$. □

Question 19. What is a forward contract? What is the forward price?

Answer: A forward contract is an agreement between two parties in which one party (the long position) agrees to buy a specified quantity of the underlying asset from the other party (the short position) at a given time in the future and for a price, called the forward price, that is agreed upon at the inception of the forward contract. The forward price is chosen such that the forward contract has value 0 at inception.[11]

If the underlying asset has spot price S_0 and pays dividends continuously at rate q, the forward price for a for-

[11] Note that the forward price is not the price of the forward.

ward contract maturing at T is

$$F = S_0 e^{(r-q)T},$$

where r is the risk–free rate between 0 and T. □

Question 20. What is the forward price for treasury futures contracts? What is the forward price for commodities futures contracts?

Answer: A short position in a forward contract (i.e., selling a forward contract) is perfectly hedged by buying one unit of the underlying asset and holding that position until the maturity of the forward contract. The forward price is the future value at the maturity of the forward contract of the cost of buying one unit of the underlying asset.

For a treasury futures contract, buying the underlying treasury bond generates a positive cash flow from receiving all the bond coupon payments until the maturity of the forward contract. If S_0 is the spot price of the underlying treasury bond and if C is the present value of all the coupon payments received during the life of the futures contract, then the forward price F of a treasury futures contract with maturity T is the future value at time T of $S_0 - C$, i.e., $F = (S_0 - C)e^{rT}$.

For a commodities futures contract, buying the underlying commodity incurs storage costs; for example, for a gold futures contract, buying the underlying gold would require storing the gold safely. If S_0 is the spot price of the underlying commodity and if C is the present value of all the storage costs, then the forward price F of a commodities futures contract with maturity T is the future value at time T of $S_0 + C$, i.e., $F = (S_0 + C)e^{rT}$, where r denotes the risk–free rate between 0 and T. □

Question 21. What is a Eurodollar futures contract?

Answer: A Eurodollar is a dollar deposit in a bank outside the United States. The Eurodollar rate is the interest rate earned on Eurodollars deposited by one bank with another bank (and is very close to LIBOR for short term maturities).

A Eurodollar futures contract is a futures contract on a Eurodollar rate, and deliveries are for up to ten years in the future.

For example, a three-month Eurodollar futures contract is a futures contract on the three month (90-day) Eurodollar rate. The start date is the third Wednesday of the delivery month (March, June,
September, December). □

Question 22. What are the most important differences between forward contracts and futures contracts?

Answer: The main differences between the ways forward and futures contracts are structured, settled, and traded are:

• Futures contracts trade on exchanges and have standard features, while forward contracts are over-the-counter instruments.

• Futures are marked to market and settled in a margin account on a daily basis, while forward contracts are not settled before maturity.

• Futures carry almost no credit and counterparty risk, since they are settled daily, while entering into a forward contract carries some credit risk.

• Futures have a range of delivery dates, while forward contracts have a specified delivery date.

• Futures contracts require the delivery of the underlying asset for the futures price, while forward contracts can be settled in cash at maturity, without the delivery of a physical asset. □

Question 23. What is the ten–day 99% VaR of a portfolio with a five–day 98% VaR of \$10 million?

Answer: If we assume normally distributed short term portfolio returns, the VaR (Value at Risk) of a portfolio is proportional to the square root of the time horizon. More precisely, if the N day $C\%$ VaR of the portfolio is denoted by $\mathrm{VaR}(N, C)$, where N is the number of days in the time horizon and C is the confidence level, then

$$\mathrm{VaR}(N, C) \approx \sigma_V z_C \sqrt{\frac{N}{252}} V(0), \qquad (3.113)$$

where, σ_V is the (annualized) standard deviation of the rate of return of the portfolio, z_C is the z–score of the standard normal distribution corresponding to C, i.e., $P(Z \leq z_C) = C$, and $V(0)$ is the current value of the portfolio.

From (3.113), it follows that

$$\mathrm{VaR}(10 \text{ days}, 99\%) \approx \frac{z_{99}\sqrt{10}}{z_{98}\sqrt{5}} \mathrm{VaR}(5 \text{ days}, 98\%).$$

For $\mathrm{VaR}(5 \text{ days}, 98\%) = \$10,000,000$, and since $z_{99} \approx 2.326348$ and $z_{98} \approx 2.053749$, we obtain that

$$\mathrm{VaR}(10 \text{ days}, 99\%) \approx \$16,019,255. \quad \square$$

Question 24. Put options with strikes 30 and 20 on the same underlying asset and with the same maturity are trading for \$6 and \$4, respectively. Can you find an arbitrage?

Answer: Since the value of a put option with strike 0 is \$0, we in fact know the prices of put options with three different strikes, i.e.,

$$P(30) = 6; \;\; P(20) = 4; \;\; P(0) = 0,$$

where $P(K)$ denotes the value of a put option with strike K.

In the plane $(K, P(K))$, these option values correspond to the points $(30, 6)$, $(20, 4)$, and $(0, 0)$, which are on the line $P(K) = \frac{2}{3}K$.

This contradicts the fact that put options are strictly convex functions of strike price, and creates an arbitrage opportunity.

The arbitrage comes from the fact that the put with strike 20 is overpriced. Using a "buy low, sell high" strategy, we could buy (i.e., go long) $\frac{2}{3}$ put options with strike 30, and sell (i.e., go short) 1 put option with strike 20. To avoid fractions, we set up the following portfolio:

- long 2 puts with strike 30;
- short 3 puts with strike 20.

This portfolio is set up at no initial cost, since the cash flow generated by selling 3 puts with strike 20 and buying 2 puts with strike 30 is \$0:

$$3 \cdot \$4 - 2 \cdot \$6 \; = \; \$0.$$

At the maturity T of the options, the value of the portfolio is

$$V(T) \; = \; 2\max(30 - S(T), 0) - 3\max(20 - S(T), 0).$$

Note that $V(T)$ is nonnegative for any value $S(T)$ of the underlying asset:

If $S(T) \geq 30$, then both put options expire worthless, and $V(T) = 0$.

If $20 \leq S(T) < 30$, then

$$V(T) \; = \; 2(30 - S(T)) \; > \; 0.$$

If $0 < S(T) < 20$, then

$$\begin{aligned} V(T) &= 2(30 - S(T)) - 3(20 - S(T)) \\ &= S(T) \\ &> 0. \end{aligned}$$

In other words, we took advantage of the existing arbitrage opportunity by setting up, at no initial cost, a portfolio with nonnegative payoff at T regardless of the price $S(T)$ of the underlying asset, and with a strictly positive payoff if $0 < S(T) < 30$. □

Question 25. I sell a one month put option with 28% implied volatility today and I hedge my position "continuously" until maturity. In one month, I calculate that the realized volatility of the underlying asset was 16%. Did I make money or did I lose money?

Answer: Delta–hedging an option "continuously"[12] is equivalent to taking an opposite position in an option with the same strike and maturity but with a volatility parameter equal to the realized volatility of the underlying asset.

Thus, the short put position which was created by selling the put at a price corresponding to 28% implied volatility is hedged by synthesizing a long put position equivalent to buying the put at a price corresponding to 16% volatility. Since option prices are increasing functions of volatility, you essentially sold a 28%–vol put and hedged it by buying the less expensive 16%–vol put, meaning that you made money on this trade. □

Question 26. Consider the following option replication strategy for a call option with strike 30 on an underlying asset with spot price \$25: If the price of the asset goes above \$30, buy one unit of the asset for \$30 and hold it while the price is above \$30. If the price of the asset goes back below \$30, sell the one unit of the asset for \$30. Thus, at maturity, you will either hold no position,

[12] In practice, the rebalancing of the hedge is done discretely (i.e., not continuously) once the rebalancing threshold is triggered, which creates additional risks related to jumps in the price of the underlying asset. Also, this hedging setup does not account for trading costs.

if the price of the asset is below the strike price 30 or you will have one unit of the asset which you bought for \$30, corresponding to the payoff of a call option with strike \$30. Seemingly, you replicated the call option at no cost. What is wrong with this argument?

Answer: We first clarify how this can be viewed as a replicating strategy for a call option with strike 30:

• if the price of the underlying asset never goes above \$30, you never buy the asset and end up with a \$0 position, which is the same as the payoff $\max(S(T)-30,0)$ of a call option with strike 30 when $S(T) < 30$; here, T denotes the maturity of the option and $S(T)$ the spot price of the underlying asset at maturity;

• every time the price of the underlying asset goes above \$30 and then back below \$30, you end up with a \$0 position, since you bought the asset for \$30 when the price went above \$30 and then sold the asset for \$30 when the price went below \$30;

• if after the last time before maturity when the asset price crosses \$30 the price ends up above \$30, the last trade you made before maturity was buying one unit of the asset for \$30; the value of your position at T is $S(T) - 30$ which is the same as the payoff $\max(S(T)-30,0)$ of the call option with strike 30 since $S(T) \geq 30$;

• if after the last time before maturity when the asset price crosses \$30 the price ends up below \$30, the last trade you made before maturity was selling one unit of the asset for \$30, which cancels out the prior buying of the asset for \$30; you end up with a \$0 position, which is the same as the payoff $\max(S(T) - 30,0)$ of a call option with strike 30 since $S(T) < 30$.

Thus, the outcome of this strategy is $\max(S(T)-30,0)$, which is exactly the payoff of a call option with strike 30 on the asset, and this was apparently achieved at no cost.

However, this cannot be the case.

The argument contains several flaws as detailed below:

Firstly, assets do not trade at one price, but rather with a bid–ask spread, where the bid price is the price at which you can sell the asset and the ask price is the price at which you can buy the asset, with the bid price always smaller than the ask price.

When the asset price goes above \$30, what you will be looking for in order to buy the asset (at a time denoted by t_1) is for \$30 to be below the bid price, i.e., $30 < S_{bid}(t_1)$. You will then buy the asset at the ask price $S_{ask}(t_1)$, which will be even higher than \$30 since $30 < S_{bid}(t_1) < S_{ask}(t_1)$. When the asset price goes below \$30, what you will be looking for in order to sell the asset (at a time denoted by t_2) is for the ask price to go below \$30, i.e., $S_{ask}(t_2) < 30$. You will then sell the asset at the bid price $S_{bid}(t_2)$, which will be even lower than \$30 since $S_{bid}(t_2) < S_{ask}(t_2) < 30$. Thus, every time the price of the underlying asset goes above \$30 at time t_1 and then back below \$30 at time t_2 you do not end up with a \$0 position from buying the asset at \$30 and selling it at \$30, but will lose money since you buy the asset at the price $S_{ask}(t_1) > 30$ and you sell the asset at the price $S_{bid}(t_2) < 30$, losing an amount equal to $S_{ask}(t_1) - S_{bid}(t_2)$.

If the price moves smoothly the loss could be small, but the loss could be significant if price jumps of the underlying asset occur when the \$30 threshold is breached. Also, the more often the \$30 price is crossed, the higher the trading losses you will incur.

Moreover, every trade incurs trading costs which were not included in the idealized trading version that seemed to generate a zero–cost replication strategy for the call option. □

Question 27. An at–the–money straddle is made of a long at–the–money call and a long at–the–money put, both with strike equal to the spot price of the underlying asset.

Consider a three months at–the–money straddle on an underlying asset with 20% volatility. Assuming that the risk–free rate and the dividend yield are zero, what is the approximate value of the delta of the straddle?

Answer: We can see intuitively that the delta of the at–the–money (ATM) straddle is approximately zero. If V denotes the value of the ATM straddle and C and P are the values of the ATM call and put, then

$$V = C + P$$

and

$$\Delta(V) = \Delta(C) + \Delta(P). \tag{3.114}$$

Recall that, for a non–dividend–paying asset,

$$\Delta(P) = \Delta(C) - 1. \tag{3.115}$$

From (3.114) and (3.115), we obtain that

$$\Delta(V) = 2\Delta(C) - 1. \tag{3.116}$$

The delta of an ATM call is about 0.5, and therefore the delta of the ATM straddle is about 0.

We are going to obtain a more precise approximation in the Black–Scholes framework involving the maturity of the option and the volatility of the underlying asset, and show that the delta of the ATM straddle is approximately 0.02.

The Black–Scholes delta of a call option is

$$\Delta(C) = e^{-qT} N(d_1), \tag{3.117}$$

where

$$d_1 = \frac{\ln\left(\frac{S}{K}\right) + \left(r - q + \frac{\sigma^2}{2}\right) T}{\sigma\sqrt{T}}; \tag{3.118}$$

here, S is the spot price of the underlying asset, T is the maturity of the option, r is the risk–free interest rate, σ is the volatility of the underlying asset, q is the continuous dividend rate of the underlying asset, and

$$N(x) = \frac{1}{\sqrt{2\pi}} \int_{-\infty}^{x} e^{-\frac{t^2}{2}}\, dt \qquad (3.119)$$

is the cumulative distribution of the standard normal variable.

For simplicity, assume that the dividend yield and the interest rates are zero, i.e., $q = 0$ and $r = 0$. Since the call option is at–the–money, the strike of the option is equal to the spot price of the underlying asset, i.e., $K = S$. Then, the formula (3.118) simplifies to

$$d_1 = \frac{\frac{\sigma^2}{2}T}{\sigma\sqrt{T}} = \frac{\sigma\sqrt{T}}{2},$$

and, from (3.117), we obtain that

$$\Delta(C) = N(d_1) = N\left(\frac{\sigma\sqrt{T}}{2}\right). \qquad (3.120)$$

We approximate the term $N\left(\frac{\sigma\sqrt{T}}{2}\right)$ from (3.120) by using a Taylor approximation around 0. Recall that

$$N(x) \approx N(0) + xN'(0), \quad \text{as } x \to 0. \qquad (3.121)$$

From (3.119), we obtain that $N'(t) = \frac{1}{\sqrt{2\pi}}e^{-\frac{t^2}{2}}$; thus, $N'(0) = \frac{1}{\sqrt{2\pi}}$, $N''(0) = 0$, Furthermore,

$$N(0) = \frac{1}{\sqrt{2\pi}} \int_{-\infty}^{0} e^{-\frac{t^2}{2}}\, dt = \frac{1}{2}.$$

Then, the Taylor approximation (3.121) becomes[13]

$$N(x) \approx \frac{1}{2} + \frac{x}{\sqrt{2\pi}}. \qquad (3.122)$$

[13]A more accurate form of the Taylor approximation (3.122) is

$$N(x) = \frac{1}{2} + \frac{x}{\sqrt{2\pi}} - \frac{x^3}{6}\frac{1}{\sqrt{2\pi}} + O(x^5), \quad \text{as } x \to 0.$$

This level of precision is not needed here.

Thus,

$$N\left(\frac{\sigma\sqrt{T}}{2}\right) \approx \frac{1}{2} + \frac{1}{\sqrt{2\pi}}\,\frac{\sigma\sqrt{T}}{2},$$

and, from (3.120), it follows that

$$\begin{aligned} \Delta(C) &\approx \frac{1}{2} + \frac{1}{\sqrt{2\pi}}\,\frac{\sigma\sqrt{T}}{2} \\ &\approx \frac{1}{2} + 0.2\sigma\sqrt{T}; \end{aligned} \tag{3.123}$$

here, we used the fact that

$$\frac{1}{\sqrt{2\pi}} = 0.39894228 \approx 0.4.$$

From (3.116) and (3.123), we obtain the following approximation for the delta of the ATM straddle:

$$\Delta(V) = 2\Delta(C) - 1 \approx 0.4\,\sigma\sqrt{T}.$$

Thus, the delta of a three months ATM straddle (i.e., with $T = \frac{3}{12} = 0.25$) on an underlying asset with 20% volatility (i.e., with $\sigma = 0.2$) is approximately

$$\Delta(V) \approx 0.4\,\sigma\sqrt{T}, = 0.02. \quad \square$$

Question 28. (i) What is the Put–Call parity for asset–or–nothing options?

(ii) What is the Put–Call parity for cash–or–nothing options?

Answer: (i) An asset–or–nothing (AoN) call option with strike K and maturity T pays the option holder the price $S(T)$ of the underlying asset at maturity if $S(T)$ is greater than the strike K, and pays nothing otherwise. In other words, the payoff at maturity of an AoN call is

$$C_{AoN}(T) = \begin{cases} S(T), & \text{if } S(T) > K; \\ 0, & \text{if } S(T) \le K. \end{cases} \tag{3.124}$$

An asset–or–nothing put option with strike K and maturity T pays the option holder the price $S(T)$ of the underlying asset at maturity if $S(T)$ is less than the strike K, and pays nothing otherwise. Thus, the payoff at maturity of an AoN put is

$$P_{AoN}(T) = \begin{cases} 0, & \text{if } S(T) \geq K; \\ S(T), & \text{if } S(T) < K. \end{cases} \qquad (3.125)$$

From (3.124) and (3.125), we find that

$$C_{AoN}(T) + P_{AoN}(T) = S(T).$$

In other words, the payoff at maturity of a long AoN call position and a long AoN put position is the same as the payoff at maturity of one unit of the underlying asset. Then, from the Generalized Law of One Price (see, e.g., Theorem 1.10 in [4]), it follows that the value at time 0 of a long AoN call position and a long AoN put position is the same as the present value at time 0 of one unit of the underlying asset at maturity. Assuming that the underlying asset pays dividends continuously at rate q, we conclude that

$$C_{AoN}(0) + P_{AoN}(0) = S(0)e^{-qT},$$

which is the Put–Call parity for asset–or–nothing options; here $S(0)$ is spot price of the underlying asset at time 0.

(ii) A cash–or–nothing (CoN) call option with strike K and maturity T pays the option holder a fixed, predetermined, cash payment B if $S(T)$ is greater than the strike K, and pays nothing otherwise. In other words, the payoff at maturity of a CoN call is

$$C_{CoN}(T) = \begin{cases} B, & \text{if } S(T) > K; \\ 0, & \text{if } S(T) \leq K. \end{cases} \qquad (3.126)$$

A cash–or–nothing put option with strike K and maturity T pays the option holder a fixed cash payment B if $S(T)$ is less than the strike K, and pays nothing otherwise. Thus, the payoff at maturity of a CoN put is

$$P_{CoN}(T) = \begin{cases} 0, & \text{if } S(T) \geq K; \\ B, & \text{if } S(T) < K. \end{cases} \quad (3.127)$$

From (3.126) and (3.127), we find that

$$C_{CoN}(T) + P_{CoN}(T) = B.$$

In other words, the payoff at maturity of a long CoN call position and a long CoN put position is the same as a long cash position B. Then, from the Generalized Law of One Price it follows that the value at time 0 of a long CoN call position and a long CoN put position is equal to the present value at time 0 of a cash position B at time T. If r denotes the risk–free rate, we conclude that

$$C_{CoN}(0) + P_{CoN}(0) = Be^{-rT},$$

which is the Put–Call parity for cash–or–nothing options. □

Question 29. Why is the price of a plain vanilla option as a function of the strike of the option a strictly convex function?

Answer: The price of a plain vanilla option as a function of the option strike is a strictly convex function for no–arbitrage reasons. In other words, if the convexity is violated, arbitrage opportunities occur.

To see this, for example, for call options, recall that the value $C(K)$ of a call option as function of the strike price K is strictly convex if and only if, for any two strikes K_1 and K_3 with $0 < K_1 < K_3$, and for any parameter λ with $0 < \lambda < 1$,

$$C(\lambda K_1 + (1-\lambda)K_3) < \lambda C(K_1) + (1-\lambda)C(K_3). \quad (3.128)$$

If convexity were violated, in other words, if there exist two strikes K_1 and K_3 with $0 < K_1 < K_3$, and a parameter λ with $0 < \lambda < 1$ such that

$$C(\lambda K_1 + (1-\lambda)K_3) \geq \lambda C(K_1) + (1-\lambda)C(K_3), \quad (3.129)$$

we will show that an arbitrage opportunity occurs.

Let

$$K_2 = \lambda K_1 + (1-\lambda)K_3. \quad (3.130)$$

Since $0 < \lambda < 1$, it follows that $K_1 < K_2 < K_3$.

Consider the following portfolio:

- long λ call options with strike K_1;
- long $1-\lambda$ call options with strike K_3;
- short one call option with strike K_2;

The setup cost[14] for this portfolio is nonnegative, since

$$\begin{aligned} & C(K_2) - \lambda C(K_1) - (1-\lambda)C(K_3) \\ = \ & C(\lambda K_1 + (1-\lambda)K_3) - \lambda C(K_1) - (1-\lambda)C(K_3) \\ \geq \ & 0; \end{aligned}$$

cf. assumption (3.129).

Let $V(T)$ be the value of the portfolio at time T, the maturity of the options. Note that $V(T)$ is nonnegative regardless of the value $S(T)$ of the underlying asset at maturity:

(i) If $S(T) \leq K_1$, then $V(T) = 0$;

(ii) If $K_1 < S(T) \leq K_2$, then

$$V(T) = \lambda(S(T) - K_1) > 0.$$

[14]Note that the set up cost for a long position with value V_0 is -\$$V_0$, since the amount \$$V_0$ is paid in order to establish the long position. Similarly, the set up cost for a short position with value V_0 is \$$V_0$, since the amount \$$V_0$ is received when establishing the short position.

(iii) If $K_2 < S(T) \leq K_3$, then

$$\begin{aligned} V(T) &= \lambda(S(T) - K_1) - (S(T) - K_2) \\ &= K_2 - \lambda K_1 - (1 - \lambda)S(T) \\ &\stackrel{(3.130)}{=} (1 - \lambda)K_3 - (1 - \lambda)S(T) \\ &\geq 0. \end{aligned}$$

(iv) If $K_3 < S(T)$, then

$$\begin{aligned} V(T) &= \lambda(S(T) - K_1) - (S(T) - K_2) \\ &\quad + (1 - \lambda)(S(T) - K_3) \\ &= K_2 - \lambda K_1 - (1 - \lambda)K_3 \\ &\stackrel{(3.130)}{=} 0. \end{aligned}$$

Thus, this portfolio generates cash when set up, has nonnegative payoff at time T regardless of the state of the market, and has positive payoffs for certain market states. According to the Generalized Law of One Price (see, e.g., Theorem 1.10 in [4]), this is an arbitrage opportunity. This is due to the fact that the option prices do not satisfy the requirement of convexity with respect to strike prices; see (3.129). Thus, in arbitrage–free markets, the prices of plain–vanilla call options must be strictly convex functions of their strikes.

Following a similar reasoning, we can show that the prices of plain vanilla put options are strictly convex functions of the strike price.

Alternatively, we can use the Put–Call Parity[15] and the already established fact that call option prices are strictly

[15]This is a widely used method: derive a relationship for call options and then use the Put–Call Parity to establish the same relationship for put options, or the other way around. For example, once the Black–Scholes value of a call option is derived, it is straightforward to use the Put–Call Parity to find the Black–Scholes value of a put option.

convex functions of their strikes to derive the convexity of put options prices as functions of strike price.

The Put–Call parity states that, for non–arbitrage, the prices $C(K)$ and $P(K)$ of a call option and of put option, respectively, with strike K and maturity T on an underlying asset with spot price S paying continuous dividends at rate q

$$P(K) = C(K) - Se^{-qT} + Ke^{-rT}, \tag{3.131}$$

where r denotes the constant risk–free rate.

We already established that, for any two strikes K_1 and K_3 with $0 < K_1 < K_3$, and a parameter λ with $0 < \lambda < 1$,

$$C(\lambda K_1 + (1-\lambda)K_3) < \lambda C(K_1) + (1-\lambda)C(K_3). \tag{3.132}$$

We will show that

$$P(\lambda K_1 + (1-\lambda)K_3) < \lambda P(K_1) + (1-\lambda)P(K_3). \tag{3.133}$$

By using (3.131) for options with strikes K_1 and K_3, we obtain that

$$\begin{aligned} P(K_1) &= C(K_1) - Se^{-qT} + K_1 e^{-rT}; \\ P(K_3) &= C(K_3) - Se^{-qT} + K_3 e^{-rT}, \end{aligned}$$

and therefore

$$\begin{aligned} & \lambda P(K_1) + (1-\lambda)P(K_3) && (3.134) \\ = \; & \lambda C(K_1) + (1-\lambda)C(K_3) && (3.135) \\ & -Se^{-qT}(\lambda K_1 + (1-\lambda)K_3)e^{-rT}. && (3.136) \end{aligned}$$

We now use (3.131) for options with strike $\lambda K_1 + (1-\lambda)K_3$ and find that

$$\begin{aligned} & P(\lambda K_1 + (1-\lambda)K_3) && (3.137) \\ = \; & C(\lambda K_1 + (1-\lambda)K_3) && (3.138) \\ & -Se^{-qT} + (\lambda K_1 + (1-\lambda)K_3)e^{-rT}. && (3.139) \end{aligned}$$

From (3.134–3.139) and using (3.132), we conclude that

$$P(\lambda K_1 + (1-\lambda)K_3) < \lambda P(K_1) + (1-\lambda)P(K_3)$$

and therefore (3.133) is established. □

Question 30. What is the sign of the ρ of a plain vanilla European option and what is the intuition behind this?

Answer: The ρ of a plain vanilla call option is positive and the ρ of a plain vanilla put option is negative; for instance, the Black–Scholes values of the ρ for a call option and for a put option are

$$\rho(C) = KTe^{-rT}N(d_2) > 0; \qquad (3.140)$$

$$\rho(P) = -KTe^{-rT}N(-d_2) < 0; \qquad (3.141)$$

here, S is the spot price of the underlying asset, T is the maturity of the option, r is the risk–free interest rate, σ is the volatility of the underlying asset, q is the continuous dividend rate of the underlying asset, and

$$d_1 = \frac{\ln\left(\frac{S}{K}\right) + \left(r - q + \frac{\sigma^2}{2}\right)T}{\sigma\sqrt{T}}; \quad d_2 = d_1 - \sigma\sqrt{T}.$$

To see this intuitively, consider a non–dividend–paying asset. In a risk–neutral world, the price of the asset grows at the risk-free rate. Then, an increase in the risk–free rate increases the forward price of the asset, thus making call option more expensive and put options cheaper. Similarly, a decrease in the risk–free rate lowers the forward price of the asset, thus making calls cheaper and puts more expensive.

Thus, call options have positive ρ and put options have negative ρ. □

Question 31. Which one is more valuable, a three–months put option with strike \$90, or a three–months call option with strike \$110, on an underlying asset with spot price \$100?

Answer: Intuitively, the out–of–the–money (OTM) call should be worth more than the OTM put. Generally speaking, the price of a European option is the present value of the expected payoff at maturity. While stock prices cannot become negative, they can, in theory, run to infinity. Thus, the upside of being long a call is unbounded, while a long put position can be worth at most the option strike at maturity.

For example, if we consider a Monte-Carlo simulation for the evolution of the price of the stock, roughly half the paths will end above \$100 and the other half below \$100. Among the paths ending above \$100, there will be paths that run up to arbitrarily high prices, but among the paths ending below \$100, there is a floor at 0. This skews the expectation of the ten percent OTM call option payoff to be higher than the payoff of the ten percent OTM put option, thus making the ten percent OTM call option more expensive than the ten percent OTM put option.

We now formally establish that a ten percent OTM call is worth more than a ten percent OTM put with the same maturity and on the same underlying asset, if the underlying asset pays no dividends and the interest rates are zero. We will then estimate that, for these specific options, the three months OTM call is worth about 50 cents more than the three months OTM put.

A ten percent OTM call with maturity T corresponds to a strike price K_C which is 10% higher than the spot price S_0 of the underlying asset, i.e., $K_C = 1.1S_0$; this is the case for our three–months call option with strike \$110 on an underlying asset with spot price \$100. A ten percent OTM put with maturity T corresponds to a strike price K_P which is 10% lower than S_0, i.e., $K_P = 0.9S_0$; this is the case for our three–months put option with strike \$90 and a spot price \$100 of the underlying asset.

In the Black–Scholes framework, with q denoting the continuous dividend rate of the underlying asset and r denoting the constant risk–free rate, the values of the options are

$$\begin{aligned} C &= S_0 e^{-qT} N(d_{1,C}) - K_C e^{-rT} N(d_{2,C}); \\ P &= K_P e^{-rT} N(-d_{2,P}) - S_0 e^{-qT} N(-d_{1,P}), \end{aligned}$$

where

$$\begin{aligned} d_{1,C} &= \frac{\ln\left(\frac{S_0}{K_C}\right) + \left(r - q + \frac{\sigma^2}{2}\right) T}{\sigma\sqrt{T}}; \\ d_{2,C} &= d_{1,C} - \sigma\sqrt{T}; \end{aligned}$$

$$\begin{aligned} d_{1,P} &= \frac{\ln\left(\frac{S_0}{K_P}\right) + \left(r - q + \frac{\sigma^2}{2}\right) T}{\sigma\sqrt{T}}; \\ d_{2,P} &= d_{1,P} - \sigma\sqrt{T}; \end{aligned}$$

here,

$$N(x) = \frac{1}{\sqrt{2\pi}} \int_{-\infty}^{x} e^{-\frac{t^2}{2}} \, dt$$

denotes the cumulative distribution of the standard normal variable.

For simplicity, assume that the asset does not pay dividends and that the interest are zero, i.e., $q = 0$ and $r = 0$. The Black–Scholes formulas become

$$\begin{aligned} C &= S_0 N(d_{1,C}) - K_C N(d_{2,C}); \\ P &= K_P N(-d_{2,P}) - S_0 N(-d_{1,P}), \end{aligned}$$

or, equivalently,

$$\begin{aligned} \frac{C}{S_0} &= N(d_{1,C}) - \frac{K_C}{S_0} N(d_{2,C}); \\ \frac{P}{S_0} &= \frac{K_P}{S_0} N(-d_{2,P}) - N(-d_{1,P}), \end{aligned}$$

where

$$d_{1,C} = \frac{\ln\left(\frac{S_0}{K_C}\right)}{\sigma\sqrt{T}} + \frac{\sigma\sqrt{T}}{2}; \quad d_{2,C} = d_{1,C} - \sigma\sqrt{T};$$
$$d_{1,P} = \frac{\ln\left(\frac{S_0}{K_P}\right)}{\sigma\sqrt{T}} + \frac{\sigma\sqrt{T}}{2}; \quad d_{2,P} = d_{1,P} - \sigma\sqrt{T}.$$

Recall that the ten percent OTM call has strike $K_C = 1.1S_0$, while the ten percent OTM put has strike $K_P = 0.9S_0$. Then, for the options,

$$\frac{C}{S_0} = N(d_{1,C}) - 1.1N(d_{2,C}); \tag{3.142}$$
$$\frac{P}{S_0} = 0.9N(-d_{2,P}) - N(-d_{1,P}), \tag{3.143}$$

with

$$d_{1,C} = \frac{\ln\left(\frac{1}{1.1}\right)}{\sigma\sqrt{T}} + \frac{\sigma\sqrt{T}}{2} = -\frac{\ln(1.1)}{\sigma\sqrt{T}} + \frac{\sigma\sqrt{T}}{2}$$
$$\approx -\frac{0.1}{\sigma\sqrt{T}} + \frac{\sigma\sqrt{T}}{2}; \tag{3.144}$$
$$d_{2,C} \approx -\frac{0.1}{\sigma\sqrt{T}} - \frac{\sigma\sqrt{T}}{2}; \tag{3.145}$$

$$d_{1,P} = \frac{\ln\left(\frac{1}{0.9}\right)}{\sigma\sqrt{T}} + \frac{\sigma\sqrt{T}}{2} = -\frac{\ln(0.9)}{\sigma\sqrt{T}} + \frac{\sigma\sqrt{T}}{2}$$
$$\approx \frac{0.1}{\sigma\sqrt{T}} + \frac{\sigma\sqrt{T}}{2}; \tag{3.146}$$
$$d_{2,P} \approx \frac{0.1}{\sigma\sqrt{T}} - \frac{\sigma\sqrt{T}}{2}; \tag{3.147}$$

here, we used the first order Taylor approximation $\ln(1+x) \approx x$ for small values of x, for $x = 0.1$ which corresponds to $\ln(1.1) \approx 0.1$, and for $x = -0.1$ which corresponds to $\ln(0.9) \approx -0.1$.

The formulas (3.142) and (3.143) require estimating the cumulative distribution function $N(x)$ of the standard normal evaluated at $d_{1,C}$, $d_{2,C}$, $d_{1,P}$, $d_{2,P}$. To do so, recall from (3.122) that the linear Taylor approximation of $N(x)$:

$$N(x) \approx \frac{1}{2} + \frac{x}{\sqrt{2\pi}}. \tag{3.148}$$

Using (3.148), we obtain from (3.142) and (3.143) that

$$\begin{aligned} \frac{C}{S_0} &\approx \frac{1}{2} + \frac{d_{1,C}}{\sqrt{2\pi}} - 1.1\left(\frac{1}{2} + \frac{d_{2,C}}{\sqrt{2\pi}}\right) \\ &= \frac{-0.1}{2} + \frac{1}{\sqrt{2\pi}}\left(d_{1,C} - 1.1 d_{2,C}\right); \end{aligned} \tag{3.149}$$

$$\begin{aligned} \frac{P}{S_0} &\approx 0.9\left(\frac{1}{2} - \frac{d_{2,P}}{\sqrt{2\pi}}\right) - \left(\frac{1}{2} - \frac{d_{1,P}}{\sqrt{2\pi}}\right) \\ &= \frac{-0.1}{2} + \frac{1}{\sqrt{2\pi}}\left(d_{1,P} - 0.9 d_{2,P}\right). \end{aligned} \tag{3.150}$$

From (3.149) and using (3.144–3.145), we find that

$$\begin{aligned} \frac{C}{S_0} &\approx -\frac{1}{20} + \frac{1}{\sqrt{2\pi}} \cdot \\ &\quad \left(-\frac{0.1}{\sigma\sqrt{T}} + \frac{\sigma\sqrt{T}}{2} - 1.1\left(-\frac{0.1}{\sigma\sqrt{T}} - \frac{\sigma\sqrt{T}}{2}\right)\right) \\ &= -\frac{1}{20} + \frac{1}{\sqrt{2\pi}}\left(\frac{0.01}{\sigma\sqrt{T}} + 1.05\sigma\sqrt{T}\right). \end{aligned} \tag{3.151}$$

From (3.150) and using (3.146–3.147), we find that

$$\begin{aligned} \frac{P}{S_0} &\approx -\frac{1}{20} + \frac{1}{\sqrt{2\pi}} \cdot \\ &\quad \left(\frac{0.1}{\sigma\sqrt{T}} + \frac{\sigma\sqrt{T}}{2} - 0.9\left(\frac{0.1}{\sigma\sqrt{T}} - \frac{\sigma\sqrt{T}}{2}\right)\right) \\ &= -\frac{1}{20} + \frac{1}{\sqrt{2\pi}}\left(\frac{0.01}{\sigma\sqrt{T}} + 0.95\sigma\sqrt{T}\right). \end{aligned} \tag{3.152}$$

From (3.151) and (3.152), we conclude that the value of a ten percent OTM call is greater than the value of a ten percent OTM put. Thus, a three–months call option with strike \$110 on an underlying asset with spot price \$100 is worth more than a three–months put option with strike \$90 on the same underlying asset.

Furthermore, from (3.151) and (3.152), it follows that

$$C - P \approx S_0 \cdot \frac{1}{\sqrt{2\pi}} 0.1\,\sigma\sqrt{T}. \qquad (3.153)$$

For $S_0 = 100$, $T = \frac{1}{4}$, $\sigma = 0.25$ (25 percent vol), and using the approximation $\sqrt{2\pi} \approx \sqrt{6.28} \approx 2.5$, we obtain from (3.153) that $C - P \approx 0.50$.

In other words, the three months OTM call is worth about 50 cents more than the three months OTM put.

Note: Thank you to Aneesh Subramanya, Baruch MFE'23, for the following clarification:

The result above was established by assuming that the implied volatility is the same for the 10% OTM call and for the 10% OTM put, or, in other words, that the volatility surface is flat, and that interest rates and dividends are zero, making the forward price is equal to spot price.

However, in the markets, the opposite is generally true: the OTM put option would be more expensive than the OTM call option.

In general, OTM puts have higher implied volatilities than OTM calls due to crash fear, and interest rates and dividends are non–zero, so the forward price is different than the spot price. The 10% OTM put would be more expensive than the 10% OTM call when moneyness is measured with respect to the forward price, not the current spot price.

Here is an example with 3–month options on the SPX index on October 24, 2024: The spot price of the index is 5810 and the forward price is 5870. The 10% OTM put has strike equal to $5870 \cdot 90\% = 5283 \approx 5280$, implied volatility equal to 21.7%, and the price of the option is \$45. The 10% OTM call has strike equal to $5870 \cdot 110\% = 6457 \approx 6450$, implied volatility equal to 11.2%, and the price of the option is \$5.50.

The 10% OTM put is worth about eight times the 10% OTM call! Typically, the difference is not this stark, but crash risk was heightened due to upcoming elections which made the implied vol skew extremely steep at that moment (21.7% for the OTM put versus 11.2% for the OTM call). □

3.4 C++. Data structures.

Question 1. How do you declare an array?

Answer: An array can be declared either on stack, or on heap.

```
//created on stack, uninitialized
T identifier[size];

//created on stack, initialized
T identifier[] = initializer_list;

//created on heap, uninitialized
T* identifier = new T[size];
```

Example:

```
int foo[3];
int bar[] = {1,2,3};
int* baz = new int[3];
```

Question 2. How do you get the address of a variable?

Answer: Use the ampersand before the name of the variable, e.g.

```
T var;
T* ptr = &var
```

Example:

```
int foo = 1;
int* foo_ptr = &foo;
```

Question 3. How do you declare an array of pointers?

Answer: The same way as declaring an array, but making the type, `T`, a pointer:

```
T* identifier[size];
T* identifier[] = initializer_list;
T** identifier = new T*[size];
```

Example:

```
int a = 1; int b = 2; int c = 3;
int* foo[3];
int* bar[] = {&a, &b, &c};
int** baz = new int*[3];
```

Question 4. How do you declare a const pointer, a pointer to a const and a const pointer to a const?

Answer:

```
//pointer to a read only variable
const T* identifier;
T const* identifier;

//read only pointer to a variable
T *const identifier = rvalue;

//read only pointer to a read only variable
const T *const identifier = rvalue;
T const *const identifier = rvalue;
```

Example:

```
//read only variables
const int a = 2; const int b = 2;
int c  = 1;

//pointer to a read only b
int const* foo_two;
foo_two = &a; foo_two = &b;

//pointer to read only a
```

```
const int* foo;
foo = &a; foo = &b;

//read only pointer to c
//it needs to be initialized
int *const bar = &c;

//read only pointer to read only a
//it needs to be initialized
const int *const baz = &a;
```

Question 5. How do you declare a dynamic array?

Answer:

```
T* identifier = new T[size];
T* identifier = nullptr;
T* identifier;

delete[] identifier;
```

Example:

```
int *foo = new int[4];
int *bar = nullptr;
bar = new int[4];
```

Question 6. What is the general form for a function signature?

Answer:

```
return_type function_name(parameter_list);
```

Example:

```
int my_sum(int a, int b);
```

Question 7. How do you pass-by-reference?

Answer:

```
return_type function_name(T & identifier);
```

The identifier is now an alias for the argument.

Question 8. How do you pass a read only argument by reference?

Answer:

```
return_type function_name(const T & identifier);
```

Once you define a parameter as `const`, you will not be able to modify it in the function.

Question 9. What are the important differences between using a pointer and a reference?

Answer: Several differences between using a pointer and a reference are:

- A pointer can be re-assigned any number of times, while a reference cannot be reassigned after initialization.
- A pointer can point to NULL (`nullptr` in C++11), while a reference can never be referred to NULL.
- It is not possible to take the address of a reference as it is done with pointers.
- There is no reference arithmetic.

Question 10. How do you set a default value for a parameter?

Answer:

```
return_type function_name(T identifier = rvalue)
```

The parameters with default value must be placed at the end of the parameter list.

Question 11. How do you create a template function?

Answer:

```
template<class T>
return_type function_name(parameter_list);

template<typename T>
return_type function_name(parameter_list);
```

Note that the parameter type can be specified, when calling the function, explicitly or implicitly. Also note that there is no technical difference between using `class` or `typename` besides code readability (`typename` for primitive types and `class` for classes).

Example:

```
template<typename T>
T temp_sum(T a, T b) {return a+b;}

struct Processor{
    int a;
    int apply(int b) {return a+b;}
};

template<class T>
int temp_sum_2(int a, int b) {
    T processor;
    processor.a = a;
    return processor.apply(b);
}

int main(){
    //implicit, foo equals 3
    int foo = temp_sum(1,2);

    //explicit, bar equals 3
    int bar = temp_sum_2<Processor>(1,2);
}
```

Question 12. How do you declare a pointer to a function?

Answer:

```
return_type (*identifier) (list_parameter_types)
```

Example:

```
int my_sum(int a, int b) {return a+b;}
int main(){
    int(*p_func)(int,int);
    p_func = & my_sum;

    // foo equals 3
    int foo = p_func(1,2);
}
```

Question 13. How do you prevent the compiler from doing an implicit conversion with your class?

Answer: Use the keyword **explicit** to define the constructor:

```
explicit Classname(parameter_list)
```

Question 14. Describe all the uses of the keyword `static` in C++.

Answer: Inside a function, using the keyword `static` means that once the variable has been initialized it remains in memory up until the end of the program.

Inside a class definition, either for a variable or for a member function, using the keyword `static` means that the there is only one copy of them per class, and shared between instances.

As a global variable in a file of code, using the keyword `static` means that the variable is private within the scope of the file.

Question 15. Can a static member function be const?

Answer: When the `const` qualifier is applied to a non-static member function it implies that member function can not change the instance class when called (i.e. can not change any non `mutable` members from `*this`). Since static member function are defined at a class level, where there is no notion of `this` the `const` qualifier for member functions does not apply.

Question 16. C++ constructors support the initialization of member data via an initializer list. When is this preferable to initialization inside the body of the constructor?

Answer: The initialization list has to be used for const members, references and with members without default constructors, but for any type of members initialization through the initialization list is still preferable, since it is for efficient. Using the initialization list, the members are initialized calling directly their constructors. If the initialization is done in the body of the constructor for each member being initialized there is an instance of it created and then a copy assignment operation is called to assign that instance to its respective member.

Question 17. What is a copy constructor, and how can the default copy constructor cause problems when you have pointer data members?

Answer: A copy constructor allows you to create a new object as a copy of an existing instance. The default copy constructor creates the new object by copying the existing object, member by member, and thus when there are member pointer you end up with two objects pointing to the same object.

It is important to note that the copy constructor is called every time a function receives an object via the pass-by-value mechanism. This means that the copy constructor needs to be implemented using a pass by reference. Otherwise you will be recursively calling the copy constructor. You should always set the parameter for a copy constructor to be const.

```
ClassName( const ClassName& other );
ClassName( ClassName& other );
ClassName( volatile const ClassName& other);
ClassName( volatile ClassName& other );
```

Question 18. What is the output of the following code:

```
#include <iostream>
using namespace std;

class A
{
public:
    int * ptr;
    ~A()
    {
        delete(ptr);
    }
};

void foo(A object_input)
{
    ;
}

int main()
{
    A aa;
    aa.ptr = new int(2);
    foo(aa);
```

```
    cout<<(*aa.ptr)<<endl;
    return 0;
}
```

Answer: The output of the code is an uncertain number, depending on the compiler used; for some compilers it could generate an error. The reason for this is that we do not define our own copy constructor.

When we call the foo function, the compiler will generate a default copy constructor which will shallow copy every data members defined in class A. This will lead to the result that two pointers, one in temporary object and the other in the object aa, will point to the same area in memory. When we get out the foo function, the compiler will automatically call the destructor function of the temporary object in which the pointer will be deleting and the area it points to will be free. In this situation, the pointer in aa will still point to the same area which has been free. When we try to visit the data through the pointer in aa, we will get garbage information.

Question 19. How do you overload an operator?

Answer:

```
type operator symbol (parameter_list);
```

If you define the operator outside of the class, then it will be a global operator function.

Example:

```
struct FooClass{int a;};
int operator + (FooClass lhs, FooClass rhs) {
    return lhs.a + rhs.a;
}
```

Question 20. What are smart pointers?

Answer: A smart pointer is a class built to mimic a pointer (offering dereferencing, indirection, arithmetic) that also offers extra features to simplify the usage, sharing and management of resources.

C++11 comes with three implementations of smart pointers: `shared_ptr`, `unique_ptr`, and `weak_ptr`.

Example:

```
//shared_pointer maintains
//a reference count
//when the count is zero the object
//pointed to is destroyed
std::shared_ptr<int> foo(new int(3));
std::shared_ptr<int> bar = foo;

//memory not released
//bar is still in scope
foo.reset();

//releases the memory,
//since no one is using it
bar.reset();
}
```

Question 21. What is encapsulation?

Answer: Encapsulation is the ability to expose an interface while hiding implementation. This is usually achieved through access modifiers (public, private, protected, etc.).

Question 22. What is a polymorphism?

Answer: Polymorphism is the ability for a set of classes to all be referenced through a common interface.

Question 23. What is inheritance?

Answer: Inheritance is the ability for one class to extend another through sub-classing. This is also referred to as "white-box" (the opposite of "black-box") re–use. A library can provide base classes that may be extended by the application developer.

Question 24. What is a virtual function? What is a pure virtual function and when do you use it?

Answer: Virtual functions are functions that are resolved by the compiler, at runtime, to the most derived version with the same signature. This means that if a function that was defined using a base class `Foo`, with a virtual member function `f`, is called using an instance of a sub class `FooChild`, that function is going to be dynamically binded to the implementation of the sub class (regardless that the actual code only refers to the base class).

A pure virtual function is a virtual function with no implementation in the base class, making the base class abstract (and thus can't be instantiated). Derived classes are forced to override the pure virtual function if they want to be instantiated. You use the same syntax as the virtual function but add `=0` to its declaration within the class.

Question 25. Why are virtual functions used for destructors? Can they be used for constructors?

Answer: Destructors are recommended to be defined as virtual, so the proper destructor (in the class hierarchy) is called at running time.

When calling a constructor, the caller needs to know the exact type of the object to be created, and thus they cannot be virtual.

Question 26. Write a function that computes the factorial of a positive integer.

Answer:

```
//for implementation
int factorial(int n){
    int output =1;
    for (int i =2 ; i <= n ; ++i)
        output *= i;
    return output;
}

//recursive implementation
int factorial(int n){
   if (n == 0) return 1;
   return n*factorial(n-1);
}

//tail recursive implementation
int factorial(int n, int last = 1){
    if (n == 0) return last;
    return factorial(n-1, last * n);
}
```

Question 27. Write a function that takes an array and returns the subarray with the largest sum.

Answer:

```
#include <vector>
#include <algorithm> // std::max

using namespace std;

template <typename T>
T max_sub_array(vector<T> const & numbers){
    T max_ending = 0, max_so_far = 0;
    for(auto & number: numbers){
        max_ending = max(0, max_ending + number);
```

```
        max_so_far = max(max_so_far, max_ending);
    }
    return max_so_far;
}
```

Question 28. Write a function that returns the prime factors of a positive integer.

Answer:

```
#include <vector>
using namepsace std;

vector<int> prime_factors(int n){
    vector<int> factors;
    for (int i = 2; i <= n/i ; ++i)
        while (n % i == 0) {
            factors.push_back(i);
            n /= i;
        }
    if (n > 1)
        factors.push_back(n);
    return factors;
}
```

Question 29. Write a function that takes a 64-bit integer and swaps the bits at indices `i` and `j`.

Answer:

```
long swap_bits (long x, const int &i, const int &j){
    if ( ((x >>i) & 1L) != ((x>>j) & 1L) )
        x ^= (1L <<i ) | (1L <<j);
    return x;
}
```

Question 30. Write a function that reverses a single linked list.

Answer:

```
#include <memory> // shared_ptr
using namespace std;

template<typename T>
struct node_t {
    T data;
    shared_ptr<node_T<T>> next;
};

//recursive implementation
template<typename T> shared_ptr<node_t<T>>
    reverse_linked_list(
        const shared_ptr<node_t <T>> &head){
    if (!head || !head->next) {
        return head;
    }

    shared_ptr<node_t<T>>
      new_head = reverse_linked_list(head->next);
    head->next->next = head;
    head->next = nullptr;
    return new_head;
}

//while implementation
template<typename T> shared_ptr<node_t<T>>
    reverse_linked_list(
        const shared_ptr<node_t <T>> &head){
    shared_ptr<node_t<T>>
        prev = nullptr, curr = head;
     while(curr) {
  shared_ptr<node_t<T>> temp = curr-> next;
  curr->next = prev;
  prev = curr;
  curr = temp;
```

```
    }
    return prev;
}
```

Question 31. Write a function that takes a string and returns true if its parenthesis are balanced.

Answer:

```
#include<string>
#include<stack>
using namespace std;

bool is_par_balanced(const string input)
{
    //"()))())"=> false
    //"(a(dd)()(()))"=>true
    stack<char> par_stack;
    for(auto &c: input)
    {
        if(c==')')
        {
            if(par_stack.empty())
            return false;
            else if(par_stack.top()=='(')
            par_stack.pop();
        }
        else if(c=='(')
        par_stack.push(c);
    }
    return par_stack.empty();
}
```

Question 32. Write a function that returns the height of an arbitrary binary tree.

Answer:

```
#include<memory> //std::shared_ptr
#include<algorithm> //std::max
using namespace std;

template <typename T>
struct BinaryTree {
    T data;
    shared_ptr<BinaryTree<T>> left, right;
};


template <typename T>
int height(
    const shared_ptr<BinaryTree<T>> &tree,
    int count = -1){
        if (!tree) return count;
        return max(
           height(tree->left, count + 1),
           height(tree->right, count +1));
}
```

Question 33. Write a C++ function that computes the n-th Fibonacci number.

Answer: The Fibonacci numbers $(F_n)_{n\geq 0}$ are given by the following recurrence:

$$F_{n+2} = F_{n+1} + F_n, \quad \forall\, n \geq 0,$$

with $F_0 = 0$ and $F_1 = 1$.

```
//recursive implementation
int fib(int n) {
    if (n == 0 || n == 1) return n;
    else {
        return fib(n-1) + fib(n-2);
```

```
    }
}

//iterative implementation
int fib(int n ){
    if (n == 0 || n == 1) return n;
    int prev = 0, last = 1, temp;
    for (int i = 2; i <= n; ++i) {
        temp = last;
        last = prev + last;
        prev = temp;
    }
    return last;
}

//tail recursive implementation
int fib(int n, int last = 1, int prev = 0)
{
    if (n == 0) return prev;
    if (n == 1) return last;
        return fib(n-1, last+prev, last);
}
```

Question 34. Implement a basic calculator to evaluate a simple expression string. The expression string may contain open parentheses "(" and closing parentheses ")", the plus sign "+" or the minus sign "-", non-negative integers and empty spaces.

Note: You may assume that the given expression is always valid. Do not use the "eval" built-in library function.

Example 1:

```
Input: "1 + 1"
Output: 2
```

Example 2:

```
Input: "2-1 + 2"
Output: 3
```

Example 3:

```
Input: "(1+(4+5+2)-3)+(6+8)"
Output: 23
```

Answer: Use a stack data structure as follows (sample code in C++):

```
class Solution {
public:
  int calculate(string s) {
    // two stacks, one on numbers, the other on operators
    stack<int> nums, ops;
    int res = 0, tmp = 0, sign = 1;
    for (int i = 0; i < s.size(); ++i) {
      if (isdigit(s[i])) {
        // parsing number (could be more than one digit)
        tmp = tmp * 10 + (s[i] - '0');
      }
      else {
        // now we are at a non-numeric char.
        // if it is addition / subtraction,
        //   we are fine to keep accumulate
        // if it is opening bracket,
        //   we save the current result
        // if it is closing bracket,
        //   we clear out previously saved op and number
        res += sign * tmp;
        tmp = 0;
        if (s[i] == '+') sign = 1;
        if (s[i] == '-') sign = -1;
        if (s[i] == '(') {
          nums.push(res);
          ops.push(sign);
          res = 0;
          sign = 1;
        }
        if (s[i] == ')' && ops.size()) {
          res = ops.top() * res + nums.top();
          ops.pop();
          nums.pop();
        }
      }
```

```
    }
    res += sign * tmp;
    return res;
  }
};
```

3.5 Statistics. Machine Learning.

Question 1. Let X_i be i.i.d uniform random variables on $[-1, 1]$. Define the random variable

$$X = \frac{\sum_{i=1}^{n} X_i}{\sqrt{\sum_{i=1}^{n} X_i^2}}.$$

Calculate the probability of $X > 1$, as $n \to \infty$.

Answer: Since X_i is a uniform random variables $U(a, b)$ with $a = -1$ and $b = 1$, we know that $\mathbb{E}[X_i] = \frac{b+a}{2} = 0$, and $\mathrm{Var}(X_i) = \mathbb{E}\left[X_i^2\right] = \frac{(b-a)^2}{12} = \frac{1}{3}$, for every i.

The random variable X can be written as

$$X = \frac{\sqrt{3} \cdot \frac{1}{\sqrt{n}} \sum_{i=1}^{n} X_i}{\sqrt{3} \cdot \sqrt{\frac{1}{n} \sum_{i=1}^{n} X_i^2}}.$$

The law of large numbers, paired with the continuous mapping theorem, yields the convergence of the denominator in probability, as $n \to \infty$:

$$\sqrt{\frac{1}{n} \sum_{i=1}^{n} X_i^2} \xrightarrow{p} \sqrt{\mathbb{E}\left[X_i^2\right]} = \frac{1}{\sqrt{3}}.$$

On the other hand, the central limit theorem yields the convergence of the numerator in distribution, as $n \to \infty$:

$$\sqrt{3} \cdot \frac{1}{\sqrt{n}} \sum_{i=1}^{n} X_i \xrightarrow{d} Z,$$

where $Z \sim N(0, 1)$ denotes the standard normal random variable.

Finally, by combining these two convergence results via Slutsky's theorem, we obtain the convergence of X in distribution, as $n \to \infty$:

$$X \xrightarrow{d} Z.$$

Therefore, for large enough n, we have

$$\mathbb{P}(X > 1) \approx \mathbb{P}(Z > 1) = 1 - N(1) \approx 0.1587,$$

where $N(\cdot)$ denotes the cdf of the standard normal distribution. □

Question 2. You are given a set of three standard 6-sided dice A, B, and C. The following table describes the outcomes of six rolls of these dices.

A	B	C
4	3	5
4	6	5
2	3	2
4	3	1
1	4	1
1	3	1

According to the data, which of them is the "most rigged"? Why?

Answer: We will calculate the Kullback-Leibler (KL) divergence for each die with respect to the fair die. The die that has the highest divergence is the one that is the "most rigged".

We will use the following definition of the KL divergence for two probability distributions $\mathbb{P}_{\mathbb{X}}$ and $\mathbb{P}_{\mathbb{Y}}$ of the discrete random variables $X, Y : \Omega \to S$:[16]

$$D\left(\mathbb{P}_{\mathbb{X}} || \mathbb{P}_{\mathbb{Y}}\right) = \sum_{s \in S} \mathbb{P}_{\mathbb{X}}(s) \log_2 \left(\frac{\mathbb{P}_{\mathbb{X}}(s)}{\mathbb{P}_{\mathbb{Y}}(s)} \right).$$

Based on the outcomes of six rolls of the dice A, B, and C, we approximate each of their three distributions

[16] We assumed that $\mathbb{P}_{\mathbb{Y}}(s) = 0$ implies $\mathbb{P}_{\mathbb{X}}(s) = 0$ for all $s \in S$, which is the case here, since the impossible events are the same for both the fair and the rigged dice.

as:

$$\begin{aligned}
\mathbb{P}_{\mathbb{A}}(3) = \mathbb{P}_{\mathbb{A}}(5) = \mathbb{P}_{\mathbb{A}}(6) = 0,\\
\mathbb{P}_{\mathbb{A}}(1) = \frac{1}{3}, \mathbb{P}_{\mathbb{A}}(2) = \frac{1}{6}, \mathbb{P}_{\mathbb{A}}(4) = \frac{1}{2};\\
\mathbb{P}_{\mathbb{B}}(1) = \mathbb{P}_{\mathbb{B}}(2) = \mathbb{P}_{\mathbb{B}}(5) = 0,\\
\mathbb{P}_{\mathbb{B}}(3) = \frac{2}{3}, \mathbb{P}_{\mathbb{B}}(4) = \mathbb{P}_{\mathbb{B}}(6) = \frac{1}{6};\\
\mathbb{P}_{\mathbb{C}}(3) = \mathbb{P}_{\mathbb{C}}(4) = \mathbb{P}_{\mathbb{C}}(6) = 0,\\
\mathbb{P}_{\mathbb{C}}(1) = \frac{1}{2}, \mathbb{P}_{\mathbb{C}}(2) = \frac{1}{6}, \mathbb{P}_{\mathbb{C}}(5) = \frac{1}{3}.
\end{aligned}$$

Furthermore, for a fair die X, we have that $\mathbb{P}_{\mathbb{X}}(i) = \frac{1}{6}$, for all $i = 1:6$.

Next, we calculate the KL divergence of each die with respect to the fair die X:

$$\begin{aligned}
& D\left(\mathbb{P}_{\mathbb{A}} || \mathbb{P}_{\mathbb{X}}\right)\\
= \; & \frac{1}{3}\log_2\left(\frac{6}{3}\right) + \frac{1}{6}\log_2\left(\frac{6}{6}\right) + \frac{1}{2}\log_2\left(\frac{6}{2}\right)\\
\approx \; & 1.13.
\end{aligned}$$

$$\begin{aligned}
& D\left(\mathbb{P}_{\mathbb{B}} || \mathbb{P}_{\mathbb{X}}\right)\\
= \; & \frac{2}{3}\log_2\left(\frac{12}{3}\right) + \frac{1}{6}\log_2\left(\frac{6}{6}\right) + \frac{1}{6}\log_2\left(\frac{6}{6}\right)\\
\approx \; & 1.33.
\end{aligned}$$

$$\begin{aligned}
& D\left(\mathbb{P}_{\mathbb{C}} || \mathbb{P}_{\mathbb{X}}\right)\\
= \; & \frac{1}{2}\log_2\left(\frac{6}{2}\right) + \frac{1}{6}\log_2\left(\frac{6}{6}\right) + \frac{1}{3}\log_2\left(\frac{6}{3}\right)\\
\approx \; & 1.13.
\end{aligned}$$

We conclude that the die B is the "most rigged," since it has the highest KL divergence. □

Question 3. You have an unfair coin which lands on heads 60% of the time. How many coin tosses are needed to detect that the coin is unfair?

Answer: The total number of heads in n coin tosses where the probability of the coin landing heads is p follows the binomial distribution $B(n,p)$, which can be approximated by the normal distribution $N(np, np(1-p))$. Thus, if the coin were fair, i.e., for $p = \frac{1}{2}$, the distribution of the total number of heads can be approximated by $N\left(\frac{n}{2}, \frac{n}{4}\right)$. To detect whether the coin is unfair, we use the Z-statistic

$$Z = \frac{X - \frac{n}{2}}{\sqrt{\frac{n}{4}}},$$

that follows the standard normal distribution $N(0,1)$. We reject the null hypothesis $H_0 : p = \frac{1}{2}$ when $\mid Z \mid > 1.96 = z_{0.05}$ at 5% significance level. Since the number of heads in n tosses is expected to be $0.6n$, we find that

$$Z = \frac{0.6n - \frac{n}{2}}{\sqrt{\frac{n}{4}}} > 1.96$$

holds when $n > 96.04$, that is, when $n \geq 97$.

We conclude that at least 97 coin tosses are needed to detect the coin is unfair at the 5% significance level. □

Question 4. Let $X_1, X_2, \ldots, X_n$ be an independent, identically distributed samples from the Gamma distribution $\Gamma\left(1, \frac{1}{\lambda}\right)$.

(a) Find the probability distribution function of the first order statistic $Y_1 = X_{(1)}$.

(b) What constant c makes cY_1 an unbiased estimator for λ?

(c) Find the MLE for λ. What is the *exact* sampling distribution of the MLE $\hat{\lambda}$? What is the bias of $\hat{\lambda}$? Find the *exact* variance of $\hat{\lambda}$.

(d) Find the Fisher information and state the asymptotic variance property for the MLE $\hat{\lambda}$ from the part (c). Compare the unbiased estimators found in parts (b) and (c). Is there any other unbiased estimator of λ better than $\hat{\lambda}$?

Answer:

(a) The Gamma distribution with the shape parameter $\alpha = 1$ and the inverse scale parameter $\beta = \frac{1}{\lambda}$ is actually an exponential distribution with the rate parameter $\frac{1}{\lambda}$. The probability density function (pdf) of X_i is thus

$$f_{X_i}(x) = \frac{1}{\lambda} e^{-\frac{x}{\lambda}}.$$

The cumulative distribution function (cdf) of X_i is

$$F_{X_i}(x) = \mathbb{P}(X \leq x) = 1 - e^{-\frac{x}{\lambda}}.$$

Therefore,

$$\begin{aligned} F_{X_{(1)}}(x) &= \mathbb{P}\left(X_{(1)} \leq x\right) = 1 - \mathbb{P}\left(X_{(1)} \geq x\right) \\ &= 1 - \mathbb{P}\left(\min_{i=1,\ldots,n} X_i \geq x\right) \\ &= 1 - \prod_{i=1}^{n} \mathbb{P}(X_i \geq x) \\ &= 1 - \prod_{i=1}^{n} e^{-\frac{x}{\lambda}} \\ &= 1 - e^{-\frac{nx}{\lambda}}. \end{aligned}$$

We conclude that the first order statistic $Y_1 = X_{(1)}$ has an exponential distribution with the rate parameter $\frac{n}{\lambda}$. The pdf of Y_1 is

$$f_{Y_1}(x) = \frac{n}{\lambda} e^{-\frac{nx}{\lambda}}.$$

(b) In order to make cY_1 an unbiased estimator for λ, we have to choose c so that $\mathbb{E}[cY_1] = \lambda$ holds. From part (a) we know that Y_1 has an exponential distribution with the rate parameter $\frac{n}{\lambda}$, so

$$\mathbb{E}[Y_1] = \frac{\lambda}{n}.$$

Therefore, we choose $c = n$, since

$$\mathbb{E}[cY_1] = \mathbb{E}[nY_1] = n \cdot \frac{\lambda}{n} = \lambda.$$

(c) The likelihood function is given by

$$L(\lambda) = \prod_{i=1}^{n} f_{X_i}(x) = \frac{1}{\lambda^n} \exp\left(-\frac{\sum_{i=1}^{n} X_i}{\lambda}\right).$$

Then, the log-likelihood function is

$$\ell(\lambda) = \log(L(\lambda)) = -n \log \lambda - \frac{\sum_{i=1}^{n} X_i}{\lambda}.$$

By taking the derivative of $\ell(\lambda)$ with respect to λ and setting it equal to zero, i.e.,

$$\frac{\partial \ell(\lambda)}{\partial \lambda} = -\frac{n}{\lambda} + \frac{\sum_{i=1}^{n} X_i}{\lambda^2} = 0,$$

we obtain that the maximum likelihood estimator (MLE) for λ is the sample mean

$$\hat{\lambda} = \frac{\sum_{i=1}^{n} X_i}{n}.$$

Recall that, if X_i, $i = 1, \ldots, n$, are independent random variables from the Gamma distributions $\Gamma(\alpha_i, \beta)$ with shape parameters α_i, respectively, and with the same inverse scale parameter β, then $\sum_{i=1}^{n} X_i$ is also Gamma–distributed with shape parameter $\sum_{i=1}^{n} \alpha_i$ and inverse scale parameter β. Using this fact, as well as the scaling property of Gamma distribution, we obtain that the exact sampling distribution of the MLE $\hat{\lambda} = \frac{\sum_{i=1}^{n} X_i}{n}$ is the

Gamma distribution with shape parameter $\sum_{i=1}^{n} 1 = n$ and scale parameter $\frac{1}{\lambda}/\frac{1}{n} = \frac{n}{\lambda}$, i.e.,

$$\hat{\lambda} \sim \Gamma\left(n, \frac{n}{\lambda}\right).$$

For $X \sim \Gamma(\alpha, \beta)$, the mean and variance of X are $\mathbb{E}[X] = \frac{\alpha}{\beta}$ and $\sigma_X^2 = \frac{\alpha}{\beta^2}$. Therefore,

$$\mathbb{E}\left[\hat{\lambda}\right] = \frac{n}{\frac{n}{\lambda}} = \lambda,$$

so $\hat{\lambda}$ is an unbiased estimator for λ. Furthermore, the variance of $\hat{\lambda}$ is

$$\sigma_{\hat{\lambda}}^2 = \frac{n}{\frac{n^2}{\lambda^2}} = \frac{\lambda^2}{n}.$$

(d) The Fisher information is given by

$$I(\lambda_0) = -\mathbb{E}_{\lambda_0}\left[\left(\frac{\partial^2}{\partial\lambda^2}\ell_\lambda(X)\right)\Big|_{\lambda=\lambda_0}\right],$$

where λ_0 is the true value of the unknown parameter λ, $f_\lambda(x) = \frac{1}{\lambda}e^{-\frac{x}{\lambda}}$, and $\ell_\lambda(x) = \log f_\lambda(x)$. Hence,

$$\begin{aligned}
\ell_\lambda(x) &= -\frac{x}{\lambda} - \log\lambda; \\
\frac{\partial}{\partial\lambda}\ell_\lambda(x) &= \frac{x}{\lambda^2} - \frac{1}{\lambda}; \\
\frac{\partial^2}{\partial\lambda^2}\ell_\lambda(x) &= -\frac{2x}{\lambda^3} + \frac{1}{\lambda^2}.
\end{aligned}$$

Finally, using the fact that $\mathbb{E}_{\lambda_0}[X] = \lambda_0$, the Fisher information of an exponentially distributed random variable X with rate parameter $\frac{1}{\lambda_0}$ is

$$I(\lambda_0) = -\mathbb{E}_{\lambda_0}\left[\left(-\frac{2X}{\lambda^3} + \frac{1}{\lambda^2}\right)\Big|_{\lambda=\lambda_0}\right] = \frac{1}{\lambda_0^2}.$$

The asymptotic variance property of the MLE $\hat{\lambda}$ states that $\sqrt{n}\left(\hat{\lambda}-\lambda_0\right)$ converges in distribution to the normally distributed random variable $N\left(0,\lambda_0^2\right)$ with mean zero and variance λ_0^2. Note that the parameters agree with the mean and variance of $\hat{\lambda}$ computed in part (c). We proceed to compute the variance of the unbiased estimator cY_1 for λ from part (b). Since $Y_1 = X_{(1)}$ has an exponential distribution with the rate parameter $\frac{n}{\lambda_0}$, we obtain from part (c) that

$$\sigma_{cY_1}^2 = c^2 \cdot \sigma_{Y_1}^2 = n^2 \cdot \left(\frac{\lambda_0}{n}\right)^2 = \lambda_0^2 > \frac{\lambda_0^2}{n} = \sigma_{\hat{\lambda}}^2.$$

Therefore, $\hat{\lambda}$ is a better (unbiased) estimator for λ than cY_1. In fact, according to the Cramér–Rao lower bound inequality, there is no unbiased estimator of λ better than $\hat{\lambda}$. $\square$

Question 5. The response variable Y is regressed onto the regressor X, and the result is $Y = aX + b$. Then X and Y switch roles; that is, X as the response is regressed onto the Y as the regressor, and the result is $X = a'Y + b'$. What is the relationship between a and a'? What is the relationship between the R^2, t- and F-statistics of the two regressions?

Answer: Recall some standard notation for a simple linear regression model:

$$\begin{aligned} SS_{xx} &= \sum_{i=1}^{n}\left(X_i-\bar{X}\right)^2 = n\,\mathrm{Var}\left(X\right); \\ SS_{xy} &= \sum_{i=1}^{n}\left(X_i-\bar{X}\right)\left(Y_i-\bar{Y}\right) = n\,\mathrm{Cov}\left(X,Y\right) \\ SS_{yy} &= \sum_{i=1}^{n}\left(Y_i-\bar{Y}\right)^2 = n\,\mathrm{Var}\left(Y\right), \end{aligned}$$

where $\bar{X}$ and $\bar{Y}$ denote the means of X_i's and Y_i's, respectively, and SS_{yy} is also known as the total sum of squares (TSS). Then,

$$a = \frac{\text{Cov}(X,Y)}{\text{Var}(X)} = \frac{SS_{xy}}{SS_{xx}}. \tag{3.154}$$

Similarly,

$$a' = \frac{\text{Cov}(X,Y)}{\text{Var}(Y)} = \frac{SS_{xy}}{SS_{yy}},$$

and therefore,

$$a \cdot a' = \left(\frac{\text{Cov}(X,Y)}{\sigma_X \cdot \sigma_Y}\right)^2 = \rho_{X,Y}^2 \leq 1,$$

where $\rho_{X,Y}$ is the correlation coefficient. Furthermore, note that $a \cdot a' = 1$ when X and Y have a perfectly linear relationship, and

$$\frac{a}{a'} = \frac{\text{Var}(Y)}{\text{Var}(X)}.$$

Denote by R_{xy}^2 and R_{yx}^2 the coefficients of determination for the simple linear regression models $Y = aX + b$ and $X = a'Y + b'$, respectively. Then,

$$R_{xy}^2 = 1 - \frac{SSR_{xy}}{SS_{yy}}, \tag{3.155}$$

where

$$SSR_{xy} = \sum_{i=1}^n e_i^2 = \sum_{i=1}^n \left(Y_i - \hat{Y}_i\right)^2$$

denotes the sum of the squared residuals for the regression model $Y = aX + b$, with

$$\hat{Y}_i = aX_i + b = aX_i + (\bar{Y} - a\bar{X}) = \bar{Y} + a(X_i - \bar{X}).$$

Note that

$$\begin{aligned} SSR_{xy} &= \sum_{i=1}^{n} \left(Y_i - \bar{Y} - a(X_i - \bar{X})\right)^2 \\ &= \sum_{i=1}^{n} \left(Y_i - \bar{Y}\right)^2 + a^2 \sum_{i=1}^{n} \left(X_i - \bar{X}\right)^2 \\ &\quad -2a \sum_{i=1}^{n} \left(Y_i - \bar{Y}\right)\left(X_i - \bar{X}\right) \\ &= SS_{yy} + a^2 SS_{xx} - 2a\, SS_{xy}. \end{aligned} \tag{3.156}$$

Since $a = \frac{SS_{xy}}{SS_{xx}}$, see (3.154), we obtain from (3.156) that

$$\begin{aligned} SSR_{xy} &= SS_{yy} - \frac{SS_{xy}^2}{SS_{xx}} \\ &= SS_{yy}\left(1 - \frac{SS_{xy}^2}{SS_{xx}SS_{yy}}\right) \\ &= SS_{yy}\left(1 - \rho_{X,Y}^2\right). \end{aligned} \tag{3.157}$$

From (3.155) and (3.157), we find that

$$R_{xy}^2 = 1 - \frac{SS_{yy}\left(1 - \rho_{X,Y}^2\right)}{SS_{yy}} = \rho_{X,Y}^2.$$

Thus, we recovered the well-known fact that, in the simple linear regression, the coefficient of determination R^2 matches the correlation coefficient squared. Since the correlation coefficient is symmetric with respect to X and Y, it follows that the coefficients of determination R_{xy}^2 and R_{yx}^2 for the simple linear regression models $Y = aX + b$ and $X = a'Y + b'$ are equal.

Recall that, in a linear regression setting, the F-statistic can be expressed as follows:

$$F = \frac{(TSS - SSR)/(p-1)}{SSR/(n-p)}, \tag{3.158}$$

where TSS denotes the total sum of squares, SSR denotes the sum of squared residuals, p is the number of predictors in the model including the intercept, and n is the number of observations. In the context of our simple linear regression models, $p = 2$ and (3.158) becomes

$$F = (n-2)\left(\frac{TSS}{SSR} - 1\right). \tag{3.159}$$

Since $R^2 = 1 - \frac{SSR}{TSS}$, we find that

$$\frac{TSS}{SSR} = \frac{1}{1-R^2}. \tag{3.160}$$

From (3.159) and (3.160), we obtain the following relationship between the R^2 and the F-statistic in any simple linear regression model:

$$F = \frac{(n-2)R^2}{1-R^2}.$$

Since we previously established that $R^2_{xy} = R^2_{yx}$ for the two simple linear regression models $Y = aX + b$ and $X = a'Y + b'$, it follows that the F-statistics for the two models are also equal.

Finally, recall that the F-statistic in a simple linear regression model is equal to the square of the t-statistic. Since the F-statistics for the two models are equal, then the t-statistics for the two models must also be equal. □

Question 6. Run the OLS regression of Y onto X_1; say R^2 is 0.4. Then, run the OLS regression of Y onto X_2; say R^2 is 0.5. Next, run the OLS regression of Y onto X_1 *and* X_2. What is the range of possible values for R^2 now?

Answer: First, notice that the R^2 for the linear regression model $Y = \beta_0 + \beta_1 X_1 + \beta_2 X_2$ cannot be lower than the R^2 for the linear regression model $Y = \beta_0 + \beta_1 X_1$ or the R^2 for the linear regression model $Y = \beta_0 + \beta_2 X_2$, since the model with both regressors could always give either $\beta_1 = 0$ or $\beta_2 = 0$ and achieve the R^2 of the other simple linear regression model. Thus, the lower bound is the maximum of the R^2's of the two simple linear regression models, which, in this case, is $\max\{0.4, 0.5\} = 0.5$.

In addition, it could very well happen that the two variables X_1 and X_2 perfectly explain Y, for example, it could be that $Y = X_1 + X_2$ is true, in which case, the upper bound of 1 on R^2 could be achieved.

We conclude that the R^2 for the linear regression model $Y = \beta_0 + \beta_1 X_1 + \beta_2 X_2$ is between 0.5 and 1. $\square$

Question 7. Compute the Kendall's τ for the Clayton copula.

Answer: The Clayton copula is part of a larger family of Archimedean copulas for which Kendall's τ can be evaluated directly from the generator $\phi(t)$ of the copula as follows:

$$\tau \;=\; 1 + 4\int_0^1 \frac{\phi(t)}{\phi'(t)}\,dt. \qquad (3.161)$$

In the case of Clayton copula,

$$\phi(t) \;=\; (t^{-\theta} - 1)/\theta,$$

see, e.g., Corollary 5.1.4 from [3], and therefore

$$\phi'(t) \;=\; -t^{-\theta-1}.$$

Then, we obtain from (3.161) that

$$\begin{aligned}
\tau &= 1 - \frac{4}{\theta}\int_0^1 \left(t - t^{\theta+1}\right)\,dt \\
&= 1 - \frac{4}{\theta}\left(\frac{1}{2}t^2 - \frac{1}{\theta+2}t^{\theta+2}\right)\Bigg|_{t=0}^{t=1} \\
&= 1 - \frac{4}{\theta}\left(\frac{1}{2} - \frac{1}{\theta+2}\right) \\
&= \frac{\theta}{\theta+2}.
\end{aligned}$$

We conclude that Kendall's τ for the Clayton copula is

$$\frac{\theta}{\theta+2}. \quad \square$$

Question 8. What are the differences between ordinary least square (OLS), ridge, LASSO, and ElasticNet regressions? How would you choose among them which method to use?

Answer: The commonality among OLS, ridge, LASSO, as well as ElasticNet regressions is that the basis functions employed in all these models are affine functions of the explanatory variables.

The differences are related to the loss functions:

- OLS attempts to determine an affine function that minimizes the loss function defined by the sum of the squared errors between the values of dependent variables (or the regressands) and those predicted by the linear model.
- Ridge regression, in addition to the same loss function as in OLS, further penalizes the loss function by a scale of the ℓ^2 norm of regression coefficients. The ℓ^2 norm penalty has two merits: (a) as in OLS, the estimators can be obtained in closed form; (b) it regularizes the problem and reduces the numerical issue of inverting the design matrix.
- LASSO, on the other hand, penalizes the loss by a scale of the ℓ^1 norm of the regression coefficients and ElasticNet by a scale of convex combination of ℓ^1 and ℓ^2 norm of the regression coefficients.
- As opposed to OLS and ridge, LASSO and ElasticNet do not have closed form expressions; one has to resort to convex optimization schemes to determine the estimators.

Although not having a closed form expression, the LASSO regression does possess a special property that ridge regression does not: as the scaling of the ℓ^1 penalty approaches infinity, estimators inferred by LASSO become zero one after another. In the case of ridge regression, as the ℓ^2 penalty approaches infinity, the estimators only fade out very slowly, in certain cases never touching zero. This property of LASSO suggests its applicability in variable selection.

As for how to choose among them, OLS provides BLUE (best linear unbiased estimator) whereas ridge and LASSO generally reduce variance although they introduce bias into the estimators due to the penalty term. □

Question 9. What is the bias-variance trade-off and how do you handle it?

Answer: Bias-variance trade-off is a broad term generally referring to the decomposition of errors related to estimations and inferences in statistics and probability.

We use two examples to demonstrate this idea:

Example 1: Assume that $\hat{\theta}$ is an estimator of the parameter θ. The mean squared error is given by

$$\mathbb{E}\left[|\hat{\theta}-\theta|^2\right].$$

Note that the mean squared error can be decomposed as

$$\begin{aligned}
&\mathbb{E}\left[|\hat{\theta}-\theta|^2\right]\\
=\ &\mathbb{E}\left[\hat{\theta}^2\right]-2\theta\mathbb{E}\left[\hat{\theta}\right]+\theta^2\\
=\ &\mathbb{E}\left[\hat{\theta}\right]^2-2\theta\mathbb{E}\left[\hat{\theta}\right]+\theta^2+\left(\mathbb{E}\left[\hat{\theta}^2\right]-\mathbb{E}\left[\hat{\theta}\right]^2\right)\\
=\ &\left(\mathbb{E}\left[\hat{\theta}\right]-\theta\right)^2+\left(\mathbb{E}\left[\hat{\theta}^2\right]-\mathbb{E}\left[\hat{\theta}\right]^2\right);
\end{aligned}$$

here, the term $\left(\mathbb{E}\left[\hat{\theta}\right]-\theta\right)^2$ is referred to as the biasedness squared and the term $\mathbb{E}\left[\hat{\theta}^2\right]-\mathbb{E}\left[\hat{\theta}\right]^2$ is the variance of the estimator $\hat{\theta}$.

While classical statistical methods focused on determining minimal variance estimators among unbiased (i.e., null biasedness) estimators, there exist examples showing that biased estimators may achieve lower errors than unbiased estimators. This indicates that sacrificing unbiaseness may help achieve even lower variance.

Example 2: When applying Monte Carlo simulations to estimate the expectation of a given random function or functional in which the state variables are driven by, for example, a stochastic differential equation (SDE), one generally has to discretize the equation and simulate finite sample paths to approximate the continuous paths in their entirety.

The estimation of the expectation of interest is thus done by evaluating the random functionals along each of the simulated paths then taking sample average. In this case, the estimation errors incur from two sources: the discretization of the underlying SDE and the finiteness of the simulated samples.

In the context of limited computational resources, the bias-variance trade-off loosely refers to the trade-off in allocating computing power to either decrease the mesh size (reducing bias) or to increase sample size (reducing variance). □

Question 10. How do you test the multicollinearity of a given set of data? How do you address the issue of significant multicollinearity?

Answer: Multicollinearity refers to an approximative collinearity or linear dependence among the explanatory variables or regressors. It makes the matrix X^tX close to singular, where X denotes the design or the exogenous matrix, resulting in numerical issues when solving linear systems corresponding to the matrix X^tX. Unless the regressors are perfectly linearly dependent, the estimators inferred by OLS are still the best linear unbiased esti-

mator (BLUE) for data with multicollinearity. However, regression coefficients can change drastically, causing the unreliability of the model. Commonly employed indices for detecting multicolinearity include:
(a) the smallest singular value and the condition number (defined as the ratio between the largest and the smallest singular values) of the design matrix;
(b) tolerance (defined as $1 - R^2$) and variance inflation factors (VIFs, defined as the reciprocal of tolerance) from auxiliary regressions.

To handle situations where multicollinearity is significant, one can attempt the following:
• perform ridge or elastic net regression to regularize the problem;
• determine and drop redundant variables by using, for example, LASSO regression;
• transform explanatory variables to mitigate the collinearity. □

Question 11. What is the loss function employed in support vector regression (SVR)?

Answer: The loss function L employed in SVR is given by

$$L(x) = \begin{cases} 0, & \text{for } |x| < \epsilon; \\ |x| - \epsilon, & \text{for } |x| \geq \epsilon, \end{cases}$$

for some $\epsilon > 0$, while for OLS the loss function is given by the quadratic function $\frac{x^2}{2}$. The loss function $L(x)$ for SVR has two main advantages over the loss function for OLS:
(a) losses from errors of magnitude smaller than ϵ are neglected, reducing chances of overfitting;
(b) losses from large errors grow linearly rather than quadratically as in OLS, increasing robustness of the resulting estimator.

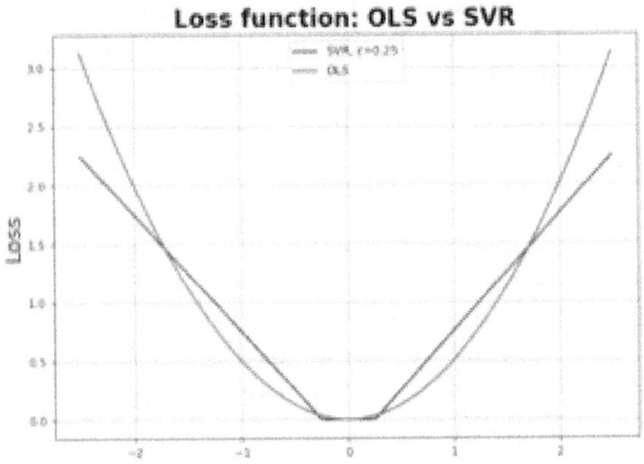

The figure above shows the loss function of SVR, with $\epsilon = 0.25$, versus that of OLS. □

3.6 Monte Carlo simulations. Numerical methods.

Question 1. How would you compute π using Monte Carlo simulations? What is the standard deviation of this method?

Answer: An Acceptance–Rejection type method can be used to approximate π as follows: generate N points uniformly in the $[-1,1] \times [-1,1]$ square and accept a point (x,y) if the point is in the unit disk $D(0,1)$, i.e., if $x^2 + y^2 \leq 1$. If the number of accepted points is A, then the ratio $\frac{A}{N}$ converges in the limit as $N \to \infty$ to the ratio of the area of the unit disk to the area of the square, which is equal to $\frac{\pi}{4}$. Therefore,

$$\pi \approx \frac{4A}{N}.$$

The standard deviation of the method is $O\left(\frac{1}{\sqrt{N}}\right)$. We make this more precise by computing below the coefficient of $\frac{1}{\sqrt{N}}$ in $O\left(\frac{1}{\sqrt{N}}\right)$; see (3.164).

Let $U_1, U_2, \cdots$ be a sequence of independent identically distributed bivariate random variables uniformly distributed in $[-1,1] \times [-1,1]$. Denote by $\mathbf{1}_{D(0,1)}$ the indicator function of the unit disk $D(0,1)$, i.e.,

$$\mathbf{1}_{D(0,1)}(x,y) = \begin{cases} 1, & \text{if } (x,y) \in D(0,1); \\ 0, & \text{otherwise.} \end{cases}$$

Let

$$X_i = \mathbf{1}_{D(0,1)}(U_i), \ \ \forall\, i \geq 1.$$

The random variables X_i, $i \geq 1$, are independent and

identically distributed, with

$$\begin{aligned} E[X_i] &= E\left[\mathbf{1}_{D(0,1)}(U_i)\right] = \iint\limits_{D(0,1)} \frac{1}{4} dx dy \\ &= \frac{\pi}{4}. \end{aligned} \tag{3.162}$$

Since X_i, $i \geq 1$, are integrable, it follows from the strong law of large numbers that

$$\lim_{n\to\infty} \frac{X_1 + X_2 + \cdots + X_n}{n} = \frac{\pi}{4} \quad \text{almost surely.}$$

Note that $X_1 + X_2 + \cdots + X_n$ counts how many points out of the randomly selected n points reside in the disk $D(0,1)$. Thus, for N large enough,

$$\frac{X_1 + X_2 + \cdots + X_N}{N} \approx \frac{\pi}{4}. \tag{3.163}$$

To calculate the variance of the estimation in (3.163), note that, for $1 \leq i \leq N$,

$$\left(\mathbf{1}_{D(0,1)}(U_i)\right)^2 = \mathbf{1}_{D(0,1)}(U_i), \quad \forall\, 1 \leq i \leq N,$$

and therefore, using (3.162), we find that

$$\begin{aligned} \mathrm{var}(X_i) &= E[X_i^2] - (E[X_i])^2 \\ &= E\left[\left(\mathbf{1}_{D(0,1)}(U_i)\right)^2\right] - \left(\frac{\pi}{4}\right)^2 \\ &= E\left[\mathbf{1}_{D(0,1)}(U_i)\right] - \frac{\pi^2}{16} \\ &= \frac{\pi}{4} - \frac{\pi^2}{16} \\ &= \frac{\pi(4-\pi)}{16}. \end{aligned}$$

Then,

$$
\begin{aligned}
& \operatorname{var}\left(\frac{X_1+X_2+\cdots+X_N}{N}\right) \\
= \quad & \frac{1}{N^2}\left(\operatorname{var}(X_1)+\cdots+\operatorname{var}(X_N)\right) \\
= \quad & \frac{1}{N^2}\cdot N\,\frac{\pi(4-\pi)}{16} \\
= \quad & \frac{\pi(4-\pi)}{16N}.
\end{aligned}
$$

By taking the square root of the variance, we conclude that the standard deviation of this Monte Carlo method for estimating π is

$$\frac{\sqrt{\pi(4-\pi)}}{4}\,\frac{1}{\sqrt{N}} \approx 0.82\,\frac{1}{\sqrt{N}}. \tag{3.164}$$

Question 2. What methods do you know for generating independent samples of the standard normal distribution?

Answer: The three most commonly used methods to generate independent standard normal samples are:

- Box–Muller (using the Marsaglia–Bray algorithm in order to avoid estimating trigonometric functions);
- Acceptance–Rejection;
- Inverse Transform.

For details on these methods, see Glasserman [2]. □

Question 3. How do you generate a geometric Brownian motion stock path using random numbers from a normal distribution?

Answer: Consider a stock whose price follows the geometric Brownian motion

$$dS_t = \mu S_t dt + \sigma S_t dW_t, \tag{3.165}$$

where μ and σ are the drift and the volatility of the stock price, and W_t is a Weiner process. To generate a price path for the stock between time 0 and time T, discretize the time interval into m equal time steps of size $\delta t = \frac{T}{m}$, and let $t_j = j\delta t$, for $j = 0 : m$.

By integrating (3.165) between t_j and t_{j+1}, it follows that

$$S_{t_{j+1}} - S_{t_j} = \mu \int_{t_j}^{t_{j+1}} S_t dt + \sigma \int_{t_j}^{t_{j+1}} S_t dW_t. \quad (3.166)$$

We use the following approximations:

$$\begin{aligned} \int_{t_j}^{t_{j+1}} S_t dt &\approx S_{t_j}(t_{j+1} - t_j) \\ &= S_{t_j}\delta t; & (3.167) \\ \int_{t_j}^{t_{j+1}} S_t dW_t &\approx S_{t_j}(W_{t_{j+1}} - W_{t_j}) \\ &= S_{t_j}\sqrt{\delta t}\, Z_{j+1}, & (3.168) \end{aligned}$$

where Z_{j+1} is a standard normal variable, since W_t is a Wiener process and therefore $W_{t_{j+1}} - W_{t_j}$ is a normal variable with mean 0 and variance $t_{j+1} - t_j = \delta t$, i.e.,

$$W_{t_{j+1}} - W_{t_j} = \sqrt{\delta t} Z_{j+1}. \quad (3.169)$$

Note that Z_j, for $j = 1 : m$, are independent standard normals.

If $z_1, z_2, \ldots, z_m$ are independent samples of the standard normal distribution, we obtain from (3.166–3.168) that (3.165) can be discretized as follows:

$$S_{t_{j+1}} - S_{t_j} = \mu S_{t_j}\delta t + \sigma S_{t_j}\sqrt{\delta t} z_{j+1},$$

for $j = 0 : (m-1)$, which can be written as

$$S_{t_{j+1}} = S_{t_j}\left(1 + \mu\delta t + \sigma\sqrt{\delta t} z_{j+1}\right),$$

for $j = 0 : (m-1)$.

Note that there is a very small probability that the price path above becomes negative, which is a drawback of using this discretization.

A price path which is always positive can be generated using Itô's formula to express (3.165) as

$$d(\ln(S_t)) = \left(\mu - \frac{\sigma^2}{2}\right) dt + \sigma dW_t. \tag{3.170}$$

By integrating (3.170) between t_j and t_{j+1}, it follows that

$$\begin{aligned} \ln(S_{t_{j+1}}) - \ln(S_{t_j}) &= \ln\left(\frac{S_{t_{j+1}}}{S_{t_j}}\right) \\ &= \left(\mu - \frac{\sigma^2}{2}\right)(t_{j+1} - t_j) + \sigma(W_{t_{j+1}} - W_{t_j}) \\ &= \left(\mu - \frac{\sigma^2}{2}\right)\delta t + \sigma\sqrt{\delta t}Z_{j+1}, \end{aligned} \tag{3.171}$$

for $j = 0 : (m-1)$, where (3.169) was used for (3.171).

Then, (3.171) can be discretized as follows:

$$\ln\left(\frac{S_{t_{j+1}}}{S_{t_j}}\right) = \left(\mu - \frac{\sigma^2}{2}\right)\delta t + \sigma\sqrt{\delta t}z_{j+1},$$

for $j = 0 : (m-1)$, and therefore

$$S_{t_{j+1}} = S_{t_j} \exp\left(\left(\mu - \frac{\sigma^2}{2}\right)\delta t + \sigma\sqrt{\delta t}z_{j+1}\right),$$

for all $j = 0 : (m-1)$. □

Question 4. How do you generate a sample of the standard normal distribution from 12 independent samples of the uniform distribution on $[0,1]$?

Answer: If $u_1, u_2, \ldots, u_{12}$ are 12 independent samples of the uniform distribution on $[0,1]$, then

$$\sum_{i=1}^{12} u_i - 6 \tag{3.172}$$

can be used as a sample of the standard normal distribution.

To see this, recall from the Central Limit Theorem that, if X_i, $i \geq 1$, are independent identically distributed random variables with finite expected value $E[X]$ and standard deviation $\sigma(X)$, then

$$\lim_{n\to\infty} \frac{\frac{1}{n}\left(\sum_{i=1}^{n} X_i\right) - E[X]}{\frac{\sigma(X)}{\sqrt{n}}} = Z, \tag{3.173}$$

where Z is the standard normal distribution and the convergence in (3.173) is in distribution.

Let U_1, U_2, ..., U_{12} be 12 independent uniform distributions on $[0, 1]$. Using (3.173) for $n = 12$ and $X_i = U_i$, $i = 1 : 12$, we infer that

$$Z \approx \frac{\frac{1}{12}\left(\sum_{i=1}^{12} U_i\right) - E[U]}{\frac{\sigma(U)}{\sqrt{12}}}, \tag{3.174}$$

where

$$E[U] = \int_0^1 u\, du = \frac{1}{2}; \tag{3.175}$$

$$\sigma^2(U) = E[U^2] - (E[U])^2 = \int_0^1 u^2 du - \left(\frac{1}{2}\right)^2 = \frac{1}{3} - \frac{1}{4} = \frac{1}{12},$$

and thus

$$\sigma(U) = \frac{1}{\sqrt{12}}. \tag{3.176}$$

From (3.174–3.176), we obtain that

$$Z \approx \frac{\frac{1}{12}\left(\sum_{i=1}^{12} U_i\right) - \frac{1}{2}}{\frac{1}{12}} = \sum_{i=1}^{12} U_i - 6,$$

and therefore (3.172) can be used as an approximate sample of the standard normal distribution.

Note that this is an inefficient method, since it uses 12 samples from the uniform distribution to generate one approximate sample of the standard normal distribution, and all the samples that it generates are in the interval $[-6, 6]$. More efficient methods for generating independent standard normal samples are Box Muller, which uses two uniform distribution samples to generate two samples of the standard normal distribution, and Acceptance–Rejection. □

Question 5. What is the rate of convergence for Monte Carlo methods?

Answer: If n is the number of paths in the Monte Carlo simulation and m is the number of time steps between 0 and T used in the discretization of each path, then the convergence rate of the Monte Carlo simulation is

$$O\left(\max\left(\frac{T}{m}, \frac{1}{\sqrt{n}}\right)\right).$$

The estimate above holds for Monte Carlo simulations on multi asset derivative securities, i.e., is independent of the number of underlying assets of the derivative security, unlike finite differences and numerical integration methods, where the convergence slows down as the number of underlying assets increases. □

Question 6. What variance reduction techniques do you know?

Answer: The variance reduction techniques are used to reduce the constant factor corresponding to the Monte Carlo approximation error $O\left(\frac{1}{\sqrt{n}}\right)$. Some of the most commonly used variance reduction techniques are:

- Control Variates;

- Antithetic Variables;
- Moment Matching.

For details on these methods and their implementation, see Glasserman [2]. □

Question 7. How do you generate samples of normal random variables with correlation ρ?

Answer: Assume that you can generate two samples z_1 and z_2 from two independent standard normal variables Z_1 and Z_2. Let $X_1 = Z_1$ and $X_2 = \rho Z_1 + \sqrt{1-\rho^2} Z_2$. Note that X_2 is a linear combination of independent normal variables, and therefore X_2 is a normal variable as well. Also,

$$\begin{aligned} \mathrm{corr}(X_1, X_2) &= \mathrm{corr}(Z_1, \rho Z_1 + \sqrt{1-\rho^2} Z_2) \\ &= \rho \,\mathrm{corr}(Z_1, Z_1) \\ &\quad + \sqrt{1-\rho^2}\,\mathrm{corr}(Z_1, Z_2) \\ &= \rho \,\mathrm{var}(Z_1) \\ &= \rho, \end{aligned}$$

since $\mathrm{var}(Z_1) = 1$ and $\mathrm{corr}(Z_1, Z_2) = 0$, since Z_1 and Z_2 are independent.

We conclude that X_1 and X_2 are normal random variables with correlation ρ. Thus, starting with two independent standard normal samples z_1 and z_2,

$$x_1 = z_1 \quad \text{and} \quad x_2 = \rho z_1 + \sqrt{1-\rho^2} z_2$$

are samples of normal random variables with correlation ρ. □

Question 8. What is the order of convergence of the Newton's method?

Answer: If it is convergent, Newton's method is quadratically (second order) convergent.

Recall that, given an initial guess x_0, the Newton's method recursion for solving $f(x) = 0$, where $f : \mathbb{R} \to \mathbb{R}$, is

$$x_{k+1} = x_k - \frac{f(x_k)}{f'(x_k)}, \quad \forall\, k \geq 0. \tag{3.177}$$

The quadratic convergence of Newton's method can be stated formally as follows: Let x^* be a solution of $f(x) = 0$. If $f(x)$ is a twice differentiable function with $f''(x)$ continuous, if $f'(x^*) \neq 0$, and if x_0 is close enough to x^*, then there exists $M > 0$ and n_M a positive integer such that

$$\frac{|x_{k+1} - x^*|}{|x_k - x^*|^2} < M, \quad \forall\, k \geq n_M. \tag{3.178}$$

To provide the intuition behind (3.178), note that, since $f(x^*) = 0$, the recursion (3.177) can be written as

$$\begin{aligned} & x_{k+1} - x^* \\ = \;& x_k - x^* - \frac{f(x_k) - f(x^*)}{f'(x_k)} \\ = \;& \frac{f(x^*) - f(x_k) + (x_k - x^*) f'(x_k)}{f'(x_k)}. \end{aligned} \tag{3.179}$$

Recall from the linear Taylor expansion of the function $f(x)$ around the point x_k that, if $f''(x)$ is continuous, then there exists a point c_k between x^* and x_k such that

$$\begin{aligned} f(x^*) &= f(x_k) + (x^* - x_k) f'(x_k) \\ &\quad + \frac{(x^* - x_k)^2}{2} f''(c_k). \end{aligned} \tag{3.180}$$

From (3.179) and (3.180), we find that

$$x_{k+1} - x^* = (x^* - x_k)^2 \, \frac{f''(c_k)}{2f'(x_k)},$$

and conclude that

$$\frac{|x_{k+1} - x^*|}{|x_k - x^*|^2} = \left| \frac{f''(c_k)}{2f'(x_k)} \right|. \tag{3.181}$$

If $f''(x)$ and $f'(x)$ are continuous functions such that $f'(x^*) \neq 0$, it follows that, if x_k is close to x^*, then $\left|\frac{f''(c_k)}{2f'(x_k)}\right|$ is close to $\left|\frac{f''(x^*)}{2f'(x^*)}\right| < \infty$. Therefore, the term $\left|\frac{f''(c_k)}{2f'(x_k)}\right|$ is uniformly bounded if x_k is close enough to x^*, and (3.181) can be written formally as (3.178). □

Question 9. Which finite difference method corresponds to trinomial trees?

Answer: Forward Euler. As an explicit finite difference method, the Forward Euler discretization of the Black–Scholes PDE gives the finite difference value of the option at a node as a linear combination of the option values at three nodes on the prior time step, which is similar to the risk neutral formula for trinomial trees.

If the calibration of the trinomial trees is done in the log space, i.e., if the up and down factors are calibrated to the normal distribution of $\ln(S)$, and if the Forward Euler discretization is done for the constant coefficients PDE obtained by the change of variables $x = \ln(S)$, the classical trinomial tree recursion and the Forward Euler recursion are almost identical. □

Question 10. What is the relationship between the LU and Cholesky decompositions?

Answer: Both the LU decomposition (without pivoting) and the Cholesky decomposition provide a computationally efficient way to solve linear systems by only using forward substitution and backward substitution, which are very fast solvers. The decompositions are similar in form, i.e., the given matrix is written as the product of a lower triangular matrix and an upper triangular matrix.

More precisely, the LU decomposition without pivoting of a nonsingular square matrix A consists of finding a lower triangular matrix L with all entries on the main diagonal equal to 1 and a nonsingular upper triangular matrix U such that

$$A = LU.$$

The matrices L and U are called the LU factors of A.

It is important to note that the LU decomposition without pivoting does not exist for every nonsingular matrix: a matrix has an LU decomposition without pivoting if and only if all the leading principal minors of the matrix are nonzero.[17] This drawback is addressed by introducing the LU decomposition with row (or column) pivoting, which exists for any nonsingular matrix.

The Cholesky decomposition of a nonsingular symmetric matrix A consists of finding a nonsingular upper triangular matrix U with positive entries on the main diagonal such that

$$A = U^t U. \tag{3.182}$$

The matrix U is called the Cholesky factor of A.

Note that not every symmetric matrix has a Cholesky decomposition: The nonsingular symmetric matrix A has a Cholesky decomposition if and only if the matrix A is symmetric positive definite, i.e., if $x^t Ax > 0$ for all $x \neq 0$, or, equivalently, if all the eigenvalues of the matrix are positive.

If they exist, both decompositions are unique. In other words, if a nonsingular matrix has an LU decomposition without pivoting, the L and U factors are uniquely determined. Similarly, any symmetric positive definite matrix has a uniquely determined Cholesky factor. □

Question 11. (i) Which matrices have an LU decomposition without pivoting?

[17]The leading principal minors of an $n \times n$ matrix A are the determinants of the $i \times i$ matrices $A_i = A(1:i, 1:i)$ made of the i^2 upper left entries of A, for $1 \leq i \leq n$.

(ii) Does a symmetric positive definite matrix have an LU decomposition without pivoting?

Answer: (i) A matrix has an LU decomposition without pivoting if and only if all the leading principal minors of the matrix are nonzero. Recall that the leading principal minors of an $n \times n$ matrix A are the determinants of the $i \times i$ matrices $A_i = A(1:i, 1:i)$ made of the i^2 upper left entries of A, for $1 \le i \le n$.[18]

Note that even very well conditioned matrices do not have an LU decomposition without pivoting. For example, the matrix $A = \begin{pmatrix} 0 & 1 \\ 1 & 1 \end{pmatrix}$ does not have an LU decomposition without row pivoting. If it did, then $A = LU$ could be written as

$$\begin{pmatrix} 1 & 0 \\ L(2,1) & 1 \end{pmatrix} \begin{pmatrix} U(1,1) & U(1,2) \\ 0 & U(2,2) \end{pmatrix} = \begin{pmatrix} 0 & 1 \\ 1 & 1 \end{pmatrix}. \tag{3.183}$$

By multiplying the matrix U by the first row of L, we find that $U(1,1) = 0$ and $U(1,2) = 1$. Thus, (3.183) becomes

$$\begin{pmatrix} 1 & 0 \\ L(2,1) & 1 \end{pmatrix} \begin{pmatrix} 0 & 1 \\ 0 & U(2,2) \end{pmatrix} = \begin{pmatrix} 0 & 1 \\ 1 & 1 \end{pmatrix}.$$

However, by multiplying the second row of L by the first column of U, it follows that

$$L(2,1) \cdot 0 + 1 \cdot 0 = 1,$$

which is not possible, and therefore we conclude that the matrix $A = \begin{pmatrix} 0 & 1 \\ 1 & 1 \end{pmatrix}$ does not have an LU decomposition without row pivoting. The reason for this is that

[18] For example, the leading principal minors of the matrix $\begin{pmatrix} 2 & -3 & 0 \\ 1 & 1 & 1 \\ -1 & 5 & -3 \end{pmatrix}$ are $\det(2) = 2$; $\det\begin{pmatrix} 2 & -3 \\ 1 & 1 \end{pmatrix} = 5$; $\det\begin{pmatrix} 2 & -3 & 0 \\ 1 & 1 & 1 \\ -1 & 5 & -3 \end{pmatrix} = -22.$

$A(1,1) = 0$, and, since $A(1,1)$ is the first leading principal minor of A, the matrix A does not have all the leading principal minors nonzero.

(ii) By definition, the $n \times n$ matrix A is symmetric positive definite (spd) if and only if $x^t Ax > 0$ for all $x \neq 0$. Several equivalent necessary and sufficient conditions for a symmetric matrix to be symmetric positive definite exist, e.g., a matrix is spd if and only if the matrix is symmetric and all the eigenvalues of the matrix are positive, and a matrix is spd if and only if the matrix is symmetric and has a Cholesky decomposition.

For the purpose of answering this question, we will use another equivalent property for spd matrices given by Sylvester's Criterion, which states that a symmetric matrix is symmetric positive definite if and only if all the leading principal minors of the matrix are positive.

Since a matrix has an LU decomposition without pivoting if and only if all the leading principal minors of the matrix are nonzero, and since the leading principal minors of a symmetric positive definite matrix are positive and therefore nonzero, we conclude that any symmetric positive definite matrix has an LU decomposition without pivoting. □

3.7 Probability. Stochastic calculus.

Question 1. What is the exponential distribution? What are the mean and the variance of the exponential distribution?

Answer: The density function of the exponential random variable X with parameter $\alpha > 0$ is

$$f(x) = \begin{cases} \alpha\, e^{-\alpha x}, & \text{if } x \geq 0; \\ 0, & \text{if } x < 0. \end{cases}$$

The expected value and the variance of the exponential random variable X are

$$E[X] = \frac{1}{\alpha} \quad \text{and} \quad \text{var}(X) = \frac{1}{\alpha^2}.$$

To see this, use integration by parts to find that

$$\begin{aligned} \int_0^\infty x e^{-\alpha x} dx &= -\left.\frac{x e^{-\alpha x}}{\alpha}\right|_0^\infty + \frac{1}{\alpha}\int_0^\infty e^{-\alpha x} dx \\ &= 0 - \left.\frac{e^{-\alpha x}}{\alpha^2}\right|_0^\infty = \frac{1}{\alpha^2}; \\ \int_0^\infty x^2 e^{-\alpha x} dx &= -\left.\frac{x^2 e^{-\alpha x}}{\alpha}\right|_0^\infty + \frac{2}{\alpha}\int_0^\infty x e^{-\alpha x} dx \\ &= 0 + \frac{2}{\alpha} \times \frac{1}{\alpha^2} = \frac{2}{\alpha^3}. \end{aligned}$$

Then,

$$E[X] = \int_{-\infty}^\infty x f(x)\, dx = \alpha \int_0^\infty x e^{-\alpha x}\, dx = \frac{1}{\alpha};$$

$$E[X^2] = \int_{-\infty}^\infty x^2 f(x)\, dx = \alpha \int_0^\infty x^2 e^{-\alpha x}\, dx = \frac{2}{\alpha^2}.$$

Therefore,

$$\text{var}(X) = E[X^2] - (E[X])^2 = \frac{2}{\alpha^2} - \left(\frac{1}{\alpha}\right)^2 = \frac{1}{\alpha^2}. \quad \square$$

Question 2. If X and Y are independent exponential random variables with mean 6 and 8, respectively, what is the probability that Y is greater than X?

Answer: The probability density functions of X and Y are, respectively:

$$f_X(x) = \begin{cases} \frac{1}{6}e^{-\frac{x}{6}}, & \text{if } x \geq 0; \\ 0, & \text{if } x < 0; \end{cases}$$

$$f_Y(y) = \begin{cases} \frac{1}{8}e^{-\frac{y}{8}}, & \text{if } y \geq 0; \\ 0, & \text{if } y < 0. \end{cases}$$

Since X and Y are independent, the joint probability density function $f_{XY}(x,y)$ of (X,Y) is the product of the marginal probability density functions, i.e.,

$$\begin{aligned} f_{XY}(x,y) &= f_X(x)\, f_Y(y) \\ &= \begin{cases} \frac{1}{48}e^{-\frac{x}{6}-\frac{y}{8}}, & \text{if } x \geq 0,\ y \geq 0; \\ 0, & \text{otherwise.} \end{cases} \end{aligned}$$

Let

$$A = \{(x,y) \in \mathbb{R}^2 : y \geq x\}.$$

The probability that Y is greater than X can be found by evaluating the double integral

$$\begin{aligned} P(Y \geq X) &= \iint_A f_{XY}(x,y)dxdy \\ &= \frac{1}{48}\int_0^\infty \int_0^y e^{-\frac{x}{6}-\frac{y}{8}}dxdy \end{aligned}$$

as follows:

$$\begin{aligned} P(Y \geq X) &= \frac{1}{48}\int_0^\infty e^{-\frac{y}{8}}\left(\int_0^y e^{-\frac{x}{6}}\,dx\right)dy \\ &= \frac{1}{48}\int_0^\infty e^{-\frac{y}{8}}\left(-6e^{-\frac{x}{6}}\Big|_0^y\right)dy \\ &= \frac{1}{8}\int_0^\infty e^{-\frac{y}{8}}\left(1 - e^{-\frac{y}{6}}\right)dy \\ &= \frac{1}{8}\int_0^\infty \left(e^{-\frac{y}{8}} - e^{-\frac{7}{24}y}\right)dy \\ &= \frac{1}{8}\left(8 - \frac{24}{7}\right) \\ &= \frac{4}{7}. \quad \square \end{aligned}$$

Question 3. What are the expected value and the variance of the Poisson distribution?

Answer: A Poisson distribution is a random variable X taking nonnegative integer values with probabilities

$$P(X = k) = \frac{e^{-\lambda}\lambda^k}{k!}, \quad \forall\, k \geq 0,$$

where $\lambda > 0$ is a fixed positive number.

We show that the expected value and the variance of the Poisson distribution X are

$$E[X] = \lambda \quad \text{and} \quad \mathrm{var}(X) = \lambda.$$

By definition,

$$\begin{aligned} E[X] &= \sum_{k=0}^\infty P(X = k)\cdot k = \sum_{k=1}^\infty \frac{e^{-\lambda}\lambda^k}{k!}\,k \\ &= e^{-\lambda}\lambda\sum_{k=1}^\infty \frac{\lambda^{k-1}}{(k-1)!}. \end{aligned} \tag{3.184}$$

Since the Taylor series expansion for e^t is

$$e^t = \sum_{k=0}^{\infty} \frac{t^k}{k!},$$

it follows that

$$\sum_{k=1}^{\infty} \frac{\lambda^{k-1}}{(k-1)!} = e^{\lambda}; \tag{3.185}$$

$$\sum_{k=2}^{\infty} \frac{\lambda^{k-2}}{(k-2)!} = e^{\lambda}. \tag{3.186}$$

From (3.184) and (3.185), we find that $E[X] = \lambda$.

To calculate $\mathrm{var}(X)$, note that

$$\begin{aligned} E[X^2] &= \sum_{k=0}^{\infty} P(X=k) \cdot k^2 \\ &= \sum_{k=1}^{\infty} \frac{e^{-\lambda}\lambda^k}{k!} k^2 \\ &= e^{-\lambda} \sum_{k=1}^{\infty} \frac{k\lambda^k}{(k-1)!}, \end{aligned}$$

which can be written as

$$\begin{aligned} E[X^2] &= e^{-\lambda} \sum_{k=1}^{\infty} \frac{(k-1)\lambda^k}{(k-1)!} + e^{-\lambda} \sum_{k=1}^{\infty} \frac{\lambda^k}{(k-1)!} \\ &= e^{-\lambda}\lambda^2 \sum_{k=2}^{\infty} \frac{\lambda^{k-2}}{(k-2)!} + e^{-\lambda}\lambda \sum_{k=1}^{\infty} \frac{\lambda^{k-1}}{(k-1)!} \\ &= \lambda^2 + \lambda, \end{aligned}$$

where (3.185) and (3.186) were used for the last equality.

We conclude that

$$\mathrm{var}(X) = E[X^2] - (E[X])^2 = \lambda. \quad \square$$

Question 4. A point is chosen uniformly from the unit disk. What is the expected value of the distance between the point and the center of the disk?

Answer: The expected value of the distance between a uniformly chosen point in the unit disk D and the center of the disk can be computed as $E\left[\sqrt{X^2+Y^2}\right]$, where (X,Y) is uniformly distributed in the unit disk D. The probability density function of (X,Y) is

$$f(x,y) = \begin{cases} \frac{1}{\pi}, & \text{if } x \in D; \\ 0, & \text{otherwise.} \end{cases}$$

Then,

$$E\left[\sqrt{X^2+Y^2}\right] = \frac{1}{\pi}\iint_D \sqrt{x^2+y^2}\,dxdy. \qquad (3.187)$$

Using the polar coordinates substitution $x = r\cos\theta$ and $y = r\sin\theta$, with $0 \le r \le 1$ and $0 \le \theta < 2\pi$, and recalling that $dxdy = rdrd\theta$, we obtain from (3.187) that

$$\begin{aligned} & E\left[\sqrt{X^2+Y^2}\right] \\ = \ & \frac{1}{\pi}\int_0^{2\pi}\int_0^1 \sqrt{r^2\left(\cos^2\theta+\sin^2\theta\right)}\,rdrd\theta \\ = \ & \frac{1}{\pi}\int_0^1\int_0^{2\pi} r^2 d\theta\, dr \qquad (3.188) \\ = \ & \frac{1}{\pi}\int_0^1 2\pi r^2\, dr \ = \ 2\int_0^1 r^2 dr \\ = \ & \frac{2}{3}, \end{aligned}$$

where for (3.188) we used the fact that $\cos^2\theta + \sin^2\theta = 1$ for all θ. □

Question 5. Consider two random variables X and Y with mean 0 and variance 1, and with joint normal distribution. If $\mathrm{cov}(X,Y) = \frac{1}{\sqrt{2}}$, what is the conditional probability $P(X>0|Y<0)$?

Answer: From the definition of conditional probability, it follows that

$$P(X>0|Y<0) = \frac{P(X>0, Y<0)}{P(Y<0)}. \tag{3.189}$$

Note that

$$P(Y<0) = \frac{1}{2}, \tag{3.190}$$

since Y is a standard normal random variable.

In order to compute $P(X>0, Y<0)$, let

$$W = \sqrt{2}\,X - Y. \tag{3.191}$$

Since $E[X] = E[Y] = 0$, it follows that $E[W] = 0$. Moreover, since

$$\mathrm{var}(X) = \mathrm{var}(Y) = 1 \quad \text{and} \quad \mathrm{cov}(X,Y) = \frac{1}{\sqrt{2}},$$

we obtain that

$$\begin{aligned}
\mathrm{var}(W) &= \mathrm{var}\left(\sqrt{2}\,X - Y\right) \\
&= \mathrm{var}\left(\sqrt{2}\,X\right) - 2\,\mathrm{cov}\left(\sqrt{2}X, Y\right) + \mathrm{var}(Y) \\
&= 2\mathrm{var}(X) - 2\sqrt{2}\,\mathrm{cov}(X,Y) + \mathrm{var}(Y) \\
&= 1,
\end{aligned}$$

and

$$\begin{aligned}
\mathrm{cov}(W,Y) &= \mathrm{cov}\left(\sqrt{2}\,X - Y, Y\right) \\
&= \sqrt{2}\,\mathrm{cov}(X,Y) - \mathrm{var}(Y) \\
&= 0.
\end{aligned}$$

Note that $W = \sqrt{2}\,X - Y$ is a normal random variable since X and Y have joint normal distribution. Moreover, since $E[W] = 0$, $\text{var}(W) = 1$, and $\text{cov}(W,Y) = 0$, it follows that W and Y are independent standard normal variables.

From (3.191), we find that

$$X = \frac{1}{\sqrt{2}}(W+Y).$$

Then, the probability of the event $\{X > 0, Y < 0\}$ can be written as

$$\begin{aligned} & P(X > 0, Y < 0) \\ = \; & P\left(\frac{1}{\sqrt{2}}(W+Y) > 0, Y < 0\right) \\ = \; & P(W+Y > 0, Y < 0). \end{aligned} \tag{3.192}$$

The two straight lines $w + y = 0$ and $y = 0$ cut the (w, y) plane into four wedges:

$$\begin{aligned} R_1 &= \{w+y > 0, y < 0\}; \\ R_2 &= \{w+y > 0, y > 0\}; \\ R_3 &= \{w+y < 0, y < 0\}; \\ R_4 &= \{w+y < 0, y > 0\}. \end{aligned}$$

Note that

$$P(W+Y > 0, Y < 0) = P((W,Y) \in R_1). \tag{3.193}$$

Since W and Y are independent normal random variables, their joint probability density function is rotationally symmetric, and therefore

$$P((W,Y) \in R_1) = P((W,Y) \in R_4); \tag{3.194}$$

$$P((W,Y) \in R_2) = P((W,Y) \in R_3); \tag{3.195}$$

$$P((W,Y) \in R_2) = 3P((W,Y) \in R_1);$$

see Figure 3.1.

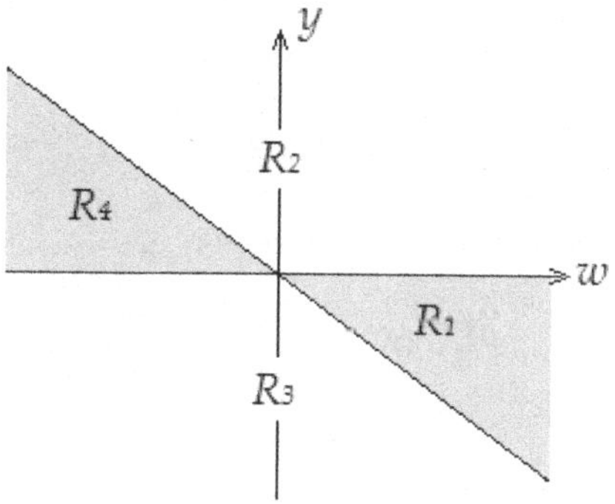

Figure 3.1: The regions R_1 to R_4 in the (w, y) plane.

Also, note that

$$\sum_{i=1}^{4} P((W, Y) \in R_i) = 1, \tag{3.196}$$

since $P(W + Y = 0 \text{ or } Y = 0) = 0$.

From (3.194–3.196), we find that

$$\begin{aligned} 1 &= \sum_{i=1}^{4} P((W, Y) \in R_i) \\ &= 2P((W, Y) \in R_1) + 2P((W, Y) \in R_2) \\ &= 8P((W, Y) \in R_1), \end{aligned}$$

and therefore

$$P((W, Y) \in R_1) = \frac{1}{8}.$$

Thus,

$$P(W + Y > 0, Y < 0) = \frac{1}{8}; \tag{3.197}$$

see (3.193).

Then, from (3.192) and (3.197), it follows that

$$P(X > 0, Y < 0) = \frac{1}{8}. \tag{3.198}$$

From (3.189), (3.190), and (3.198), we conclude that

$$\begin{aligned} P(X > 0 | Y < 0) &= \frac{P(X > 0, Y < 0)}{P(Y < 0)} \\ &= \frac{1}{4}. \quad \square \end{aligned}$$

Question 6. If X and Y are lognormal random variables, is their product XY lognormally distributed?

Answer: First, note that, if X and Y are independent lognormal random variables, then XY is lognormally distributed, since $\ln(XY) = \ln(X) + \ln(Y)$ is the sum of two independent normal random variables, and therefore it is normally distributed.

In a more general setting, if $\ln(X)$ and $\ln(Y)$ have joint normal distribution, then $\ln(X) + \ln(Y)$ is normally distributed and therefore XY is a lognormal random variable.

Otherwise, $\ln(X) + \ln(Y)$ may not be normally distributed even if $\ln(X)$ and $\ln(Y)$ are normally distributed, in which case XY is not lognormally distributed. $\square$

Question 7. Let X be a normal random variable with mean μ and variance σ^2, and let Φ be the cumulative distribution function of the standard normal distribution. Find the expected value of $Y = \Phi(X)$.

Answer: Let Z be a standard normal random variable independent of X. Then,

$$Y = \Phi(X) = P(Z \le X | X) = E[\mathbf{1}_{Z \le X} | X],$$

and therefore

$$E[Y] = E\left[E\left[\mathbf{1}_{Z \leq X} \mid X\right]\right]. \qquad (3.199)$$

Recall from the Tower Property for conditional expectation[19] that, for any two random variables T and W,

$$E[T] = E[E[T|W]]. \qquad (3.200)$$

Using (3.200) for $T = \mathbf{1}_{Z \leq X}$ and $W = X$, we obtain that

$$\begin{aligned} E\left[E\left[\mathbf{1}_{Z \leq X} \mid X\right]\right] &= E\left[\mathbf{1}_{Z \leq X}\right] \\ &= P(Z \leq X). \end{aligned} \qquad (3.201)$$

From (3.199) and (3.201), we obtain that

$$E[Y] = P(Z \leq X). \qquad (3.202)$$

Recall that X is a normal random variable with mean μ and variance σ^2, and that X and Z are independent. Then, $Z - X$ is a normal random variable with the following mean and variance:

$$\begin{aligned} E[Z - X] &= -\mu; \\ \mathrm{var}(Z - X) &= \mathrm{var}(Z) + \mathrm{var}(X) = 1 + \sigma^2. \end{aligned}$$

and $Z - X$ is a normal random variable with mean $-\mu$ and variance $1 + \sigma^2$. Thus, $\frac{Z-X+\mu}{\sqrt{1+\sigma^2}}$ is a standard normal random variable and therefore

$$\begin{aligned} P(Z \leq X) &= P(Z - X \leq 0) \\ &= P\left(\frac{Z - X + \mu}{\sqrt{1+\sigma^2}} \leq \frac{\mu}{\sqrt{1+\sigma^2}}\right) \\ &= \Phi\left(\frac{\mu}{\sqrt{1+\sigma^2}}\right). \end{aligned} \qquad (3.203)$$

[19] In other words, to calculate the expected value of T, one can first calculate the conditional expected value of T knowing the extra information from W, then average out the resulting conditional expected value over W.

From (3.202) and (3.203) we conclude that

$$E[Y] = \Phi\left(\frac{\mu}{\sqrt{1+\sigma^2}}\right). \qquad (3.204)$$

We note that, if X is the standard normal variable, then $Y = \Phi(X)$ is uniformly distributed in the interval $[0,1]$, and therefore $E[Y] = \frac{1}{2}$. Note that, for $\mu = 0$ and $\sigma = 1$ in (3.204), i.e., if X is a standard normal variable, then $E[Y] = \Phi(0) = \frac{1}{2}$, which is consistent to the comment above. □

Question 8. What is the law of large numbers?

Answer: There is a strong law of large numbers and there is a weak law of large numbers. The strong law of large numbers states that the average of a large number of independent identically distributed integrable random variables converges almost surely to their common mean; in the case of the weak law of large numbers, the convergence is only in probability.

More precisely, let $X_1, X_2, \ldots$ be a sequence of independent identically distributed random variables with finite expected value $\mu = E[X_i]$, and let $S_n = X_1 + \cdots + X_n$.

The strong law of large numbers states that $\frac{S_n}{n} \rightarrow \mu$ almost surely, i.e.,

$$P\left(\lim_{n\rightarrow\infty} \frac{S_n}{n} = \mu\right) = 1. \qquad (3.205)$$

The weak law of large numbers states that $\frac{S_n}{n} \rightarrow \mu$ in probability, i.e.,

$$\lim_{n\rightarrow\infty} P\left(\left|\frac{S_n}{n} - \mu\right| > \epsilon\right) = 0, \quad \forall\, \epsilon > 0. \qquad (3.206)$$

Note that, if a sequence of random variables convergences almost surely, then it also converges in probability, and therefore, if (3.205) holds true, then (3.206) also holds true. This is the sense in which the strong law of large numbers is "stronger" than the weak law of large numbers. □

Question 9. What is the central limit theorem?

Answer: The central limit theorem states that the limiting distribution of the centered and scaled sum of an independent identically distributed sequence of random variables is a normal distribution if the common distribution of the random variables has finite variance.

More precisely, let $X_1, X_2, \ldots$ be a sequence of independent identically distributed random variables with finite expected value $\mu = E[X_i]$ and finite variance $\sigma^2 = \text{var}[X_i]$. Let $S_n = X_1 + \cdots + X_n$. Then,

$$\lim_{n\to\infty} \frac{S_n - n\mu}{\sigma\sqrt{n}} = Z,$$

where Z is the standard normal distribution, and the convergence is in distribution, i.e.,

$$\lim_{n\to\infty} P\left(\frac{S_n - n\mu}{\sigma\sqrt{n}} \leq t\right) = P(Z \leq t).$$

Putting together the Law of Large Numbers and the Central Limit Theorem, the following approximation as n goes to infinity holds:

$$\frac{S_n}{n} \approx \mu + \frac{\sigma}{\sqrt{n}} Z. \quad \square$$

Question 10. What is a martingale? How is it related to option pricing?

Answer: Let $(\Omega, \mathcal{F}_t, P)$ be a filtered probability space, where Ω is the sample space, $\mathcal{F}_t$ is a filtration, and P is a probability measure on Ω. A stochastic process X_t

is called a *martingale* with respect to the filtration $\{\mathcal{F}_t\}$ if and only if
(i) X_t is adapted, i.e., X_t is $\mathcal{F}_t$-measurable for all t;
(ii) X_t is integrable for all t, i.e., $E[|X_t|] < \infty$ for all t;
(iii) $E[X_t|\mathcal{F}_s] = X_s$ almost surely for all $s < t$.

In other words, a martingale is a stochastic process in which, given the available information $\mathcal{F}_s$ up to the current time s, the optimal estimation (in the least square sense) of the process in the future time t, i.e., $E[X_t|\mathcal{F}_s]$, is the current value X_s almost surely.

The martingale concept is one of the cornerstones of option pricing theory. The fundamental theorem of asset pricing states that, if a market model is arbitrage–free, then there exists a risk neutral probability so that each discounted asset price process under the risk neutral probability is a martingale. Thus, a way to price a derivative security is to figure out a partial differential equation, called the pricing equation, usually deduced by applying Itô's formula, so that the discounted price process of the derivative is a martingale under risk neutral probability.

Question 11. Explain the assumption $(dW_t)^2 = dt$ used in the informal derivation of Itô's Lemma.

Answer: The notation in differential form $(dW_t)^2 = dt$ is a shorthand for the conventional integral notation used in Riemann integral

$$\int_0^T (dW_t)^2 = \int_0^T dt. \tag{3.207}$$

The intuition behind (3.207) is related to the *quadratic variation* of a Brownian path. By definition, the quadratic variation $QV_W[0,T]$ of the Brownian path W in the interval $[0,T]$ is

$$QV_W[0,T] = \lim_{n\to\infty} \sum_{i=1}^{n} \left|W_{t_i} - W_{t_{i-1}}\right|^2,$$

where $t_i = \frac{i}{n}T$, for $i = 0 : n$. Therefore, in expectation, we have that

$$\begin{aligned} & E\left[\lim_{n\to\infty}\sum_{i=1}^{n}\left|W_{t_i} - W_{t_{i-1}}\right|^2\right] \\ = & \lim_{n\to\infty}\sum_{i=1}^{n} E\left[\left|W_{t_i} - W_{t_{i-1}}\right|^2\right] \\ = & \lim_{n\to\infty}\sum_{i=1}^{n}(t_i - t_{i-1}) \\ = & T, \end{aligned}$$

since $W_{t_i} - W_{t_{i-1}} \sim N(0, t_i - t_{i-1})$ for $i = 1 : n$. In fact, it can be shown that the convergence

$$\sum_{i=1}^{n}\left|W_{t_i} - W_{t_{i-1}}\right|^2 \longrightarrow T \text{ as } n \to \infty$$

is in L^2 sense, i.e.,

$$\lim_{n\to\infty} E\left[\left|\sum_{i=1}^{n}\left|W_{t_i} - W_{t_{i-1}}\right|^2 - T\right|^2\right] = 0.$$

By mimicking the notations used in the conventional Riemann integral, we write that

$$\begin{aligned} \int_0^T (dW_t)^2 &= \lim_{n\to\infty}\sum_{i=1}^{n}\left|W_{t_i} - W_{t_{i-1}}\right|^2 = T \\ &= \int_0^T dt, \end{aligned}$$

whose shorthand notation in differential form is

$$(dW_t)^2 = dt. \quad \square$$

Question 12. If W_t is a Wiener process, find $E[W_t W_s]$.

Answer: Assume that $s \leq t$. Write W_t as $W_t = (W_t - W_s) + W_s$, and note that

$$W_t W_s \; = \; (W_t - W_s) W_s \; + \; W_s^2. \tag{3.208}$$

Since W_τ, $\tau \geq 0$, is a Wiener process, it follows that $W_t - W_s$ and W_s are independent normal random variables of mean 0, and therefore

$$\begin{aligned} E[(W_t - W_s) W_s] \; &= \; E[W_t - W_s] E[W_s] \\ &= \; 0. \end{aligned} \tag{3.209}$$

Also, W_s is normal of mean $E[W_s] = 0$ and variance $\mathrm{var}(W_s) = s$. Thus,

$$\begin{aligned} \mathrm{var}(W_s) \; &= \; E[W_s^2] \; - \; (E[W_s])^2 \\ &= \; E[W_s^2], \end{aligned}$$

and therefore

$$E[W_s^2] \; = \; \mathrm{var}(W_s) \; = \; s. \tag{3.210}$$

From (3.208–3.210), we obtain that

$$\begin{aligned} E[W_t W_s] \; &= \; E[(W_t - W_s) W_s] + E[W_s^2] \\ &= \; s. \end{aligned} \tag{3.211}$$

Since (3.211) was derived under the assumption $s \leq t$, we conclude that

$$E[W_t W_s] \; = \; \min(s, t). \quad \Box \tag{3.212}$$

Question 13. If W_t is a Wiener process, what is $\mathrm{var}(W_t + W_s)$?

Answer: Assume that $s \leq t$. Write W_t as $W_t = (W_t - W_s) + W_s$, and note that $W_t + W_s = (W_t - W_s) + 2W_s$.

Then,

$$\begin{aligned}\operatorname{var}(W_t + W_s) &= \operatorname{var}(W_t - W_s) \\ &\quad + 4\operatorname{cov}(W_t - W_s, W_s) \\ &\quad + 4\operatorname{var}(W_s). \qquad (3.213)\end{aligned}$$

Since W_τ, $\tau \geq 0$, is a Wiener process, it follows that

$$\begin{aligned}\operatorname{cov}(W_t - W_s, W_s) &= 0; \qquad (3.214)\\ \operatorname{var}(W_t - W_s) &= t - s; \qquad (3.215)\\ \operatorname{var}(W_s) &= s, \qquad (3.216)\end{aligned}$$

since $W_t - W_s$ and W_s are independent normal variables of variance $t - s$ and s, respectively.

From (3.213–3.216), we obtain that

$$\operatorname{var}(W_t + W_s) = (t - s) + 4s = t + 3s. \qquad (3.217)$$

Since (3.217) was derived under the assumption $s \leq t$, we conclude that

$$\operatorname{var}(W_t + W_s) = \max(s, t) + 3\min(s, t). \quad \square$$

Question 14. Let W_t be a Wiener process. Find Find

$$\int_0^t W_s \, dW_s \quad \text{and} \quad E\left[\int_0^t W_s \, dW_s\right].$$

Answer: Since $\int x \, dx = \frac{x^2}{2} + C$, we begin by computing $d\left(\frac{W_t^2}{2}\right)$. Recall from Itô's lemma that, if $f(x,t)$ is a continuously differentiable function, then

$$df = \frac{\partial f}{\partial t} \, dt + \frac{\partial f}{\partial x} \, dW_t + \frac{1}{2}\frac{\partial^2 f}{\partial x^2} \, dt.$$

For $f(x,t) = \frac{x^2}{2}$, we find that

$$d\left(\frac{W_t^2}{2}\right) = W_t dW_t + \frac{1}{2} dt,$$

and therefore

$$d\left(\frac{W_t^2 - t}{2}\right) = W_t dW_t. \qquad (3.218)$$

By integrating (3.218) between 0 and t, and since $W_0 = 0$, we obtain that

$$\int_0^t W_s \, dW_s = \frac{W_t^2 - t}{2}, \quad \forall\, t \geq 0. \qquad (3.219)$$

From (3.219), it follows that

$$E\left[\int_0^t W_s \, dW_s\right] = \frac{E[W_t^2] - t}{2} = 0,$$

since W_t is normally distributed with mean 0 and variance t and therefore

$$\begin{aligned} E[W_t^2] &= \operatorname{var}(W_t) + (E[W_t])^2 \\ &= t. \quad \square \end{aligned}$$

Question 15. Find the distribution of the random variable

$$X = \int_0^1 W_t dW_t.$$

Answer: Recall from Itô's formula that, if W_t is a Wiener process and $f(x)$ is a function with continuous second order derivative, then

$$df(W_t) = f'(W_t)dW_t + \frac{1}{2}f''(W_t)dt. \qquad (3.220)$$

For $f(x) = x^2$, we obtain from (3.220) that

$$dW_t^2 = 2W_t dW_t + dt. \qquad (3.221)$$

By integrating (3.221) from 0 to 1, we obtain that

$$W_1^2 = 2\int_0^1 W_t dW_t + 1 = 2X + 1. \qquad (3.222)$$

Note that W_1 is a standard normal random variable, and let $W_1 = Z$. By solving (3.222) for X, we find that

$$X = \frac{W_1^2 - 1}{2} = \frac{Z^2 - 1}{2}.$$

Let $f_X(x)$ and $F_X(x)$ be the probability density function of X and the cumulative distribution function of X, respectively.

Note that

$$P\left(X \le -\frac{1}{2}\right) = P(Z^2 \le 0) = 0.$$

Thus, if $x < -\frac{1}{2}$, then $F_X(x) = P(X \le x) = 0$ and therefore

$$f_X(x) = 0,$$

since $f_X(x) = F_X'(x)$.

If $x \ge -\frac{1}{2}$, then

$$\begin{aligned} F_X(x) &= P(X \le x) = P\left(\frac{Z^2 - 1}{2} \le x\right) \\ &= P\left(Z^2 \le 2x + 1\right) \\ &= P\left(-\sqrt{2x+1} \le Z \le \sqrt{2x+1}\right) \\ &= 2\, P\left(0 \le Z \le \sqrt{2x+1}\right) \\ &= \frac{2}{\sqrt{2\pi}} \int_0^{\sqrt{2x+1}} e^{-\frac{y^2}{2}}\, dy. \end{aligned}$$

By differentiating $F_X(x)$, it follows that, for $x < -\frac{1}{2}$,

$$\begin{aligned} f_X(x) &= F_X'(x) \\ &= \sqrt{\frac{2}{\pi}}\frac{d}{dx}\left(\int_0^{\sqrt{2x+1}} e^{-\frac{y^2}{2}}dy\right) \\ &= \sqrt{\frac{2}{\pi}}e^{-\frac{(\sqrt{2x+1})^2}{2}}\left(\sqrt{2x+1}\right)' \\ &= \sqrt{\frac{2}{\pi}}\frac{e^{-\frac{2x+1}{2}}}{\sqrt{2x+1}}. \end{aligned}$$

We conclude that

$$f_X(x) = \begin{cases} \sqrt{\frac{2}{\pi}}\frac{e^{-\frac{2x+1}{2}}}{\sqrt{2x+1}}, & \text{if } x > -\frac{1}{2}; \\ 0 & \text{if } x \le -\frac{1}{2}. \end{cases} \qquad \square$$

Question 16. Let W_t be a Wiener process. Find the mean and the variance of

$$\int_0^t W_s^2 dW_s.$$

Answer: Let $\mathcal{B}_t$ be the Borel σ-algebra over the time interval $[0,t]$ and let $\mathcal{F}_t$ be the filtration for the probability space in which the Wiener process W_t resides. We will use the following results:[20]

Theorem 3.1. *(Martingality)*
Let f_s be progressively measurable and square integrable in $[0,T]$, i.e., f_t is $\mathcal{B}_t \otimes \mathcal{F}_t$-measurable for every $t \in [0,T]$ and $E\left[\int_0^T |f_t|^2 dt\right] < \infty$. Then, the stochastic integral

[20] For the proofs of Theorem 3.1 and Theorem 3.2, see Theorem 2.8 on p.65 and Theorem 3.1 on p.67, respectively, from Friedman [1].

$\int_0^t f_s dW_s$ defines a zero mean, square integrable martingale for $t \in [0,T]$. In particular,

$$E\left[\int_0^t f_s dW_s\right] = 0, \quad \forall\, t \in [0,T]. \tag{3.223}$$

Theorem 3.2. *(Itô's isometry)*
Let f_t and g_t be progressively measurable and square integrable processes. Then,

$$E\left[\int_0^t f_s dW_s \int_0^t g_s dW_s\right] = \int_0^t E[f_s\, g_s] ds.$$

In particular, if $f_s = g_s$, it follows that

$$E\left[\left(\int_0^t f_s dW_s\right)^2\right] = \int_0^t E[f_s^2] ds. \tag{3.224}$$

For our problem, we need to compute $E[X]$ and $\mathrm{var}(X)$, where

$$X = \int_0^t W_s^2 dW_s.$$

We first check that the integrand W_s^2 is progressively measurable and square integrable in $[0,t]$.

Note that W_s^2 is progressively measurable because it is adapted and continuous.

Furthermore, since $W_s \sim N(0,s)$, it follows that $W_s = \sqrt{s}Z$, where Z is a standard normal random variable, and therefore

$$E[W_s^4] = E\left[\left(\sqrt{s}Z\right)^4\right] = s^2\, E[Z^4] = 3s^2, \tag{3.225}$$

where we used the fact that the fourth moment of the standard normal distribution is 3, i.e., $E[Z^4] = 3$. From (3.225), it follows that

$$\int_0^t E[W_s^4] ds = \int_0^t 3s^2 ds = t^3 < \infty. \tag{3.226}$$

In other words, W_s^2 is square integrable in $[0,t]$.

We can therefore apply both Theorems 3.1 and 3.2 with $f_s = W_s^2$ and $g_s = W_s^2$.

From (3.223) for $f_s = W_s^2$, we find that

$$E[X] = E\left[\int_0^t W_s^2 dW_s\right] = 0. \tag{3.227}$$

From (3.224) for $f_s = g_s = W_s^2$, and using (3.226), it follows that

$$\begin{aligned} E[X^2] &= E\left[\left(\int_0^t W_s^2 dW_s\right)^2\right] = \int_0^t E[W_s^4]ds \\ &= t^3. \end{aligned}$$

Since $E[X] = 0$, see (3.227), we find that

$$\text{var}(X) = E[X^2] - (E[X])^2 = t^3. \quad \square$$

Question 17. If W_t is a Wiener process, find the variance of

$$X = \int_0^1 \sqrt{t}e^{\frac{W_t^2}{8}} dW_t.$$

Answer: We will use Theorems 3.1 and 3.2 to solve this problem. To do so, we first check that the integrand $\sqrt{t}e^{\frac{W_t^2}{8}}$ is progressively measurable and square integrable in $[0,1]$.

The process $\sqrt{t}e^{\frac{W_t^2}{8}}$ is progressively measurable because it is adapted and continuous.

Furthermore, since $W_t \sim N(0,t)$, the probability density function of W_t is $\frac{1}{\sqrt{2\pi t}} e^{-\frac{x^2}{2t}}$, and therefore

$$\begin{aligned}
E\left[e^{\frac{W_t^2}{4}}\right] &= \frac{1}{\sqrt{2\pi t}} \int_{-\infty}^{\infty} e^{\frac{x^2}{4}} e^{-\frac{x^2}{2t}}\, dx \\
&= \frac{1}{\sqrt{2\pi t}} \int_{-\infty}^{\infty} e^{\left(\frac{1}{4}-\frac{1}{2t}\right)x^2}\, dx \\
&= \frac{1}{\sqrt{2\pi t}} \int_{-\infty}^{\infty} e^{-\frac{1}{2}\cdot\frac{2-t}{2t}x^2}\, dx \\
&= \frac{1}{\sqrt{2\pi t}} \sqrt{2\pi \cdot \frac{2t}{2-t}} \qquad (3.228) \\
&= \sqrt{\frac{2}{2-t}}, \qquad (3.229)
\end{aligned}$$

where, for (3.228), we used the identity

$$\int_{-\infty}^{\infty} e^{-\frac{Ax^2}{2}}\, dx = \sqrt{\frac{2\pi}{A}},$$

for any positive constant $A > 0$. From (3.229), it follows that

$$\begin{aligned}
\int_0^1 E\left[te^{\frac{W_t^2}{4}}\right] dt &= \int_0^1 t\, E\left[e^{\frac{W_t^2}{4}}\right] dt \\
&= \int_0^1 t \cdot \sqrt{\frac{2}{2-t}}\, dt \\
&= -\frac{2\sqrt{2}}{3} \left((t+4)\sqrt{2-t}\right)\Big|_0^1 \\
&= \frac{2}{3}\left(8 - 5\sqrt{2}\right). \qquad (3.230)
\end{aligned}$$

Thus, $\int_0^1 E\left[te^{\frac{W_t^2}{4}}\right] dt < \infty$, and we conclude that both Theorems 3.1 and 3.2 can be applied here.

From (3.223), it follows that

$$E[X] = E\left[\int_0^1 \sqrt{t} e^{\frac{W_t^2}{8}}\, dW_t\right] = 0. \qquad (3.231)$$

From (3.224), and using (3.230), we obtain that

$$\begin{aligned} E[X^2] &= E\left[\left(\int_0^1 \sqrt{t}e^{\frac{W_t^2}{8}}\,dW_t\right)^2\right] \\ &= \int_0^1 E\left[\left(\sqrt{t}e^{\frac{W_t^2}{8}}\right)^2\right] dt \\ &= \int_0^1 E\left[te^{\frac{W_t^2}{4}}\right] dt \\ &= \frac{2}{3}(8-5\sqrt{2}). \qquad (3.232) \end{aligned}$$

From (3.231) and (3.232), we conclude that

$$\mathrm{var}(X) = E[X^2] - (E[X])^2 = \frac{2}{3}\left(8-5\sqrt{2}\right). \quad \square$$

Question 18. If W_t is a Wiener process, what is $E\left[e^{W_t}\right]$?

Answer:
Solution 1: If W_t is a Wiener process and $Y = e^{W_t}$, then

$$\ln(Y) = W_t \cong \sqrt{t}Z, \qquad (3.233)$$

where Z is the standard normal random variable, and therefore Y is a lognormal random variable. Recall that the expected value of a lognormal random variable given by $\ln(V) = \mu + \sigma Z$ is

$$E[V] = e^{\mu + \frac{\sigma^2}{2}}. \qquad (3.234)$$

From (3.233), and using (3.234) with $\mu = 0$ and $\sigma = \sqrt{t}$, we conclude that

$$E\left[e^{W_t}\right] = E[Y] = e^{\frac{t}{2}}.$$

Solution 2: Since $W_t \cong \sqrt{t}Z$, it follows that

$$E\left[e^{W_t}\right] = E\left[e^{\sqrt{t}Z}\right] = \frac{1}{\sqrt{2\pi}} \int_{-\infty}^{\infty} e^{\sqrt{t}x} e^{-\frac{x^2}{2}}\, dx$$

$$= \frac{1}{\sqrt{2\pi}} \int_{-\infty}^{\infty} e^{\sqrt{t}x - \frac{x^2}{2}}\, dx. \qquad (3.235)$$

By completing the square, we find that

$$\sqrt{t}x - \frac{x^2}{2} = -\frac{1}{2}\left(x - \sqrt{t}\right)^2 + \frac{t}{2},$$

and therefore

$$e^{\sqrt{t}x - \frac{x^2}{2}} = e^{\frac{t}{2}} e^{-\frac{1}{2}(x-\sqrt{t})^2}. \qquad (3.236)$$

From (3.235) and (3.236), it follows that

$$E\left[e^{W_t}\right] = e^{\frac{t}{2}} \frac{1}{\sqrt{2\pi}} \int_{-\infty}^{\infty} e^{-\frac{1}{2}(x-\sqrt{t})^2}\, dx$$

$$= e^{\frac{t}{2}} \frac{1}{\sqrt{2\pi}} \int_{-\infty}^{\infty} e^{-\frac{y^2}{2}}\, dy \qquad (3.237)$$

$$= e^{\frac{t}{2}}, \qquad (3.238)$$

where we used the substitution $y = x - \sqrt{t}$ for (3.237), and, for (3.238), we used the fact that

$$\frac{1}{\sqrt{2\pi}} \int_{-\infty}^{\infty} e^{-\frac{y^2}{2}}\, dy = 1,$$

since $\frac{1}{\sqrt{2\pi}} e^{-\frac{y^2}{2}}$ is the probability density function of the standard normal distribution. □

Question 19. If W_t is a Wiener process, find the variance of

$$\int_0^t s\, dW_s.$$

Answer: Recall that, for any deterministic square integrable function $f : [a, b] \to \mathbb{R}$, the stochastic integral $\int_a^b f(s)dW_s$ is normally distributed with mean 0 and variance equal to the square of the L^2 norm of f, i.e.,

$$\int_a^b f(s)dW_s \sim N\left(0, \int_a^b f^2(s)ds\right).$$

Then,

$$\int_0^t s\,dW_s \;\sim\; N\left(0, \int_0^t s^2 ds\right) \;=\; N\left(0, \frac{t^3}{3}\right),$$

and we conclude that

$$\mathrm{var}\left(\int_0^t sdW_s\right) = \frac{t^3}{3}. \quad \square$$

Question 20. Let W_t be a Wiener process, and let

$$X_t \;=\; \int_0^t W_\tau d\tau. \tag{3.239}$$

What is the distribution of X_t? Is X_t a martingale?

Answer: A solution to this question was given in Chapter 1 using integration by parts; we include a different solution here.

Note that X_t is not a martingale because, if we rewrite (3.239) in differential form as

$$dX_t \;=\; W_t dt \;=\; W_t dt + 0\,dW_t,$$

we can think of X_t as being a diffusion process with only the drift part W_t.

Recall that the integral of a one-parameter family of Gaussian random variables remains Gaussian. Since W_t is a Gaussian family, X_t is normally distributed. Furthermore,

$$E[X_t] = \int_0^t E[W_\tau]d\tau = 0.$$

Therefore,

$$\mathrm{var}(X_t) = E[X_t^2] - (E[X_t])^2 = E[X_t^2]. \qquad (3.240)$$

Note that

$$X_t^2 = \left(\int_0^t W_s ds\right)^2 = \int_0^t \int_0^t W_s W_u dsdu, \qquad (3.241)$$

and recall that

$$E[W_s W_u] = \min\{s, u\}, \quad \forall\ s, u > 0; \qquad (3.242)$$

see (3.212).

From (3.240–3.242), we obtain that

$$\begin{aligned}
\mathrm{var}(X_t) &= E[X_t^2] = \int_0^t \int_0^t E[W_s W_u] dsdu \\
&= \int_0^t \int_0^t \min\{s, u\} dsdu \\
&= \int_0^t \left(\int_0^u sds + \int_u^t uds\right) du \\
&= \int_0^t \left(\frac{u^2}{2} + u(t-u)\right) du \\
&= \int_0^t \left(ut - \frac{u^2}{2}\right) du \\
&= \frac{t^3}{2} - \frac{t^3}{6} \\
&= \frac{t^3}{3}.
\end{aligned}$$

We conclude that X_t is normally distributed with mean 0 and variance $\frac{t^3}{3}$, i.e.,

$$X_t \sim N\left(0, \frac{t^3}{3}\right). \quad \square$$

Question 21. What is an Itô process?

Answer: An Itô process is a generic term referring to a stochastic process X_t determined by the solution of a stochastic differential equation (SDE) of the form

$$dX_t = a(X_t, t)dt + b(X_t, t)dW_t, \qquad (3.243)$$

where W_t is a Wiener process. The coefficient $a(x,t)$ of the dt term is the drift of X_t; the coefficient $b(x,t)$ of the dW_t term is the diffusion of X_t. The SDE (3.243) is, by definition, the shorthand notation for the stochastic integral equation

$$X_t = X_0 + \int_0^t a(X_s, s)\, ds + \int_0^t b(X_s, s)\, dW_s.$$

We note that a sufficient condition for the existence and uniqueness of the (strong) solution to an SDE is for the drift and the diffusion coefficients $a(x,t)$ and $b(x,t)$ to be locally Lipschitz functions of at most linear growth in x. □

Question 22. What is Itô's lemma?

Answer: Itô's lemma, also known as Itô's formula, states that, if X_t is an Itô process satisfying the SDE

$$dX_t = a(X_t, t)dt + b(X_t, t)dW_t,$$

then for any function $f(x,t)$ with continuous second order partial derivative in x and continuous first order partial derivative in t, the process $f(X_t, t)$ is also an Itô process, driven by the same Wiener process W_t, and the drift and diffusion parts of $f(X_t, t)$ are determined according to the Taylor expansion of $f(x,t)$ to first order for the t part and up to second order for the x part:

$$\begin{aligned} df(X_t,t) &= \frac{\partial f}{\partial t}\,dt + \frac{\partial f}{\partial x}\,dX_t + \frac{1}{2}\frac{\partial^2 f}{\partial x^2}\,(dX_t)^2 \\ &= \frac{\partial f}{\partial t}\,dt + \frac{\partial f}{\partial x}\,[a(X_t,t)dt + b(X_t,t)dW_t] \\ &\quad + \frac{1}{2}\frac{\partial^2 f}{\partial x^2}\,[a(X_t,t)dt + b(X_t,t)dW_t]^2 \\ &= \left[\frac{\partial f}{\partial t} + a(X_t,t)\frac{\partial f}{\partial x} + \frac{b^2(X_t,t)}{2}\frac{\partial^2 f}{\partial x^2}\right] dt \\ &\quad + b(X_t,t)\frac{\partial f}{\partial x}\,dW_t. \end{aligned} \tag{3.244}$$

Note that for (3.244) we used the fact that

$$\begin{aligned} (dX_t)^2 &= [a(X_t,t)dt + b(X_t,t)dW_t]^2 \\ &= b^2(X_t,t)dt \end{aligned}$$

since $(dW_t)^2 = dt$ (see Problem 3.7 in this section), $(dt)^2 = 0$, and $dW_t\,dt = 0$. □

Question 23. If W_t is a Wiener process, is the process $X_t = W_t^2$ a martingale?

Answer: Recall that a stochastic process M_t defined in a filtered probability space $(\Omega, \mathcal{F}_t, P)$ is a martingale if and only if
(i) M_t is adapted, i.e., M_t is $\mathcal{F}_t$-measurable for all t;
(ii) M_t is integrable for all t, i.e., $E[M_t] < \infty$ for all t;
(iii) $E[|M_t||\mathcal{F}_s] = M_s$ almost surely for all $s < t$.

We check whether the process X_t satisfies the conditions (i), (ii), and (iii) above.

Since any continuous function of the Wiener process W_t is adapted, the process X_t is adapted, and therefore condition (i) is satisfied.

Also, $E[X_t] = E[W_t^2] = t < \infty$, since W_t is a normal random variable of mean 0 and variance t, and therefore

$E[W_t^2] = \text{var}[W_t] = t$. Thus, the random variable X_t is integrable for every t, and X_t satisfies condition (ii).

To check whether X_t satisfies condition (iii), note that, for every $s < t$,

$$\begin{aligned} E\left[X_t|\mathcal{F}_s\right] &= E\left[W_t^2|\mathcal{F}_s\right] \\ &= E\left[(W_t - W_s + W_s)^2|\mathcal{F}_s\right] \\ &= E\left[(W_t - W_s)^2|\mathcal{F}_s\right] + 2E\left[(W_t - W_s)W_s|\mathcal{F}_s\right] \\ &\quad + E\left[W_s^2|\mathcal{F}_s\right]. \end{aligned} \tag{3.245}$$

Since the Wiener process W_t has independent increments, i.e, $W_t - W_s$ is independent of $\mathcal{F}_s$, and stationary increment, i.e., $W_t - W_s \sim W_{t-s}$. Moreover, $W_t - W_s$ is a normal random variable of mean 0 and variance $t - s$, i.e., $E[W_{t-s}] = 0$ and $\text{var}(W_{t-s}) = E[W_{t-s}^2] = t - s$. Then,

$$E\left[(W_t - W_s)^2|\mathcal{F}_s\right] = E[W_{t-s}^2] = t - s, \tag{3.246}$$

and

$$\begin{aligned} E\left[(W_t - W_s)W_s|\mathcal{F}_s\right] &= W_s E\left[W_t - W_s|\mathcal{F}_s\right] \\ &= W_s E[W_{t-s}] \\ &= 0. \end{aligned} \tag{3.247}$$

Since W_s is $\mathcal{F}_s$-measurable, it follows that

$$E[W_s^2|\mathcal{F}_s] = W_s^2. \tag{3.248}$$

From (3.245–3.248), we find that

$$\begin{aligned} &E\left[X_t|\mathcal{F}_s\right] \\ &= E\left[(W_t - W_s)^2|\mathcal{F}_s\right] + 2E\left[(W_t - W_s)W_s|\mathcal{F}_s\right] \\ &\quad + E\left[W_s^2|\mathcal{F}_s\right] \\ &= t - s + W_s^2. \end{aligned}$$

Thus,

$$E\left[X_t|\mathcal{F}_s\right] = t - s + W_s^2 \neq W_s^2 = X_s,$$

and therefore the process X_t does not satisfy condition (iii).

We conclude that X_t is not a martingale. $\square$

Question 24. If W_t is a Wiener process, is the process

$$N_t = W_t^3 - 3tW_t$$

a martingale?

Answer:

Solution 1: A stochastic process M_t defined in a filtered probability space $(\Omega, \mathcal{F}_t, P)$ is called a martingale if and only if

(i) M_t is adapted, i.e., M_t is $\mathcal{F}_t$-measurable for all t;
(ii) M_t is integrable for all t, i.e., $E[|M_t|] < \infty$ for all t;
(iii) $E[M_t|\mathcal{F}_s] = M_s$ almost surely for all $s < t$.

We check whether the process N_t satisfies the conditions (i), (ii), and (iii) above.

Since any continuous function of the Wiener process W_t is adapted, the process N_t is adapted, and therefore condition (i) is satisfied.

Also,

$$E[N_t] = E[W_t^3] - 3tE[W_t] = 0 < \infty,$$

since $W_t \sim N(0, t)$, therefore $E[W_t] = E[W_t^3] = 0$. Thus, the random variable N_t is integrable for every t, and N_t satisfies condition (ii).

To check whether N_t satisfies condition (iii), note that, for every $s < t$,

$$\begin{aligned}
& E\left[N_t|\mathcal{F}_s\right] \\
= \ & E\left[W_t^3 - 3tW_t|\mathcal{F}_s\right] \\
= \ & E\left[(W_t - W_s + W_s)^3|\mathcal{F}_s\right] - 3tE\left[W_t|\mathcal{F}_s\right] \\
= \ & E\left[(W_t - W_s)^3|\mathcal{F}_s\right] + 3E\left[(W_t - W_s)^2 W_s|\mathcal{F}_s\right] \\
& + \ 3E\left[(W_t - W_s)W_s^2|\mathcal{F}_s\right] \ + \ E\left[W_s^3|\mathcal{F}_s\right] \\
& - \ 3tE\left[W_t|\mathcal{F}_s\right].
\end{aligned}$$

Recall that the Wiener process W_t has independent increments, i.e, $W_t - W_s$ is independent of $\mathcal{F}_s$, and stationary increment, i.e., $W_t - W_s \sim W_{t-s}$, and the fact that W_s is $\mathcal{F}_s$-measurable. Then,

$$\begin{aligned} E\left[(W_t - W_s)^3 | \mathcal{F}_s\right] &= E[W_{t-s}^3] \\ &= 0; \qquad (3.249) \\ E\left[(W_t - W_s)^2 W_s | \mathcal{F}_s\right] &= W_s \; E[W_{t-s}^2] \\ &= (t-s)W_s; \qquad (3.250) \\ E\left[(W_t - W_s) W_s^2 | \mathcal{F}_s\right] &= W_s^2 \; E[W_{t-s}] \\ &= 0; \qquad (3.251) \\ E\left[W_s^3 | \mathcal{F}_s\right] &= W_s^3, \qquad (3.252) \end{aligned}$$

since $W_{t-s} \sim N(0, t-s)$, and therefore

$$E[W_{t-s}] = E[W_{t-s}^3] = 0;$$

$$E[W_{t-s}^2] = \text{var}[W_{t-s}] = t - s.$$

Moreover, since W_t is a martingale, it follows that

$$E[W_t | \mathcal{F}_s] = W_s. \qquad (3.253)$$

From (3.249–3.253), we find that

$$\begin{aligned} & E\left[N_t | \mathcal{F}_s\right] \\ = \; & E\left[(W_t - W_s)^3 | \mathcal{F}_s\right] + 3E\left[(W_t - W_s)^2 W_s | \mathcal{F}_s\right] \\ & + \; 3E\left[(W_t - W_s) W_s^2 | \mathcal{F}_s\right] \; + \; E\left[W_s^3 | \mathcal{F}_s\right] \\ & - \; 3tE\left[W_t | \mathcal{F}_s\right] \\ = \; & 0 + 3(t-s)W_s + 0 + W_s^3 - 3tW_s \\ = \; & W_s^3 - 3sW_s \\ = \; & N_s. \end{aligned}$$

Thus, the process N_t satisfies condition (iii), and we conclude that N_t is a martingale.

Solution 2: By applying Itô's formula to N_t, we obtain that

$$\begin{aligned} dN_t &= d(W_t^3 - 3tW_t) \\ &= 3W_t^2 dW_t + \frac{1}{2} \cdot 6W_t dt \;-\; 3(W_t dt + t dW_t) \\ &= 3(W_t^2 - t) dW_t. \end{aligned}$$

The process N_t has zero drift and can be written as a stochastic integral as

$$N_t = \int_0^t 3(W_s^2 - s) dW_s.$$

Moreover,

$$\begin{aligned} & \int_0^t E\left[\left(3(W_s^2 - s)\right)^2 \right] ds \\ = \;& 9 \left[\int_0^t \left(E[W_s^4] - 2sE[W_s^2] + s^2 \right) ds \right] \\ = \;& 9 \left[\int_0^t \left(3s^2 - 2s^2 + s^2 \right) ds \right] \qquad (3.254) \\ = \;& 6t^3 < \infty; \end{aligned}$$

for (3.254), we used the fact that $W_s = \sqrt{s}Z$, where Z is a standard normal random variable, and therefore

$$\begin{aligned} E[W_s^2] &= E\left[(\sqrt{s}Z)^2 \right] = sE[Z^2] = s; \\ E[W_s^4] &= E\left[(\sqrt{s}Z)^4 \right] = s^2\, E[Z^4] = 3s^2. \end{aligned}$$

Therefore, the process $3(W_s^2 - s)$ is square integrable for $s \in [0, t]$, and, from Theorem 3.1, we conclude that N_t is a martingale.

Question 25. What is Girsanov's theorem?

Answer: Girsanov's theorem is providing a way to change the drift of a Wiener process by defining a new probability measure via a Radon-Nikodym derivative. More precisely, let $(\Omega, \mathcal{F}_t, P)$, for $0 \leq t \leq T$, be a filtered probability space, and let W_t be a Wiener process in the probability measure P. Let h_t be a progressively measurable stochastic process such that the stochastic exponential

$$\mathcal{E}_t(h) = \exp\left(\int_0^t h_s dW_s - \frac{1}{2}\int_0^t h_s^2 ds\right)$$

is a martingale in the probability measure P. Define a new probability measure $\widetilde{P}$ over Ω given by the Radon-Nikodym derivative as follows:

$$\frac{d\widetilde{P}}{dP} = \exp\left(\int_0^T h_t dW_t - \frac{1}{2}\int_0^T h_t^2 dt\right). \tag{3.255}$$

Then, W_t is a Wiener process with drift h under the new probability measure $\widetilde{P}$. Equivalently, if we define $\widetilde{W}_t$ by $\widetilde{W}_t = W_t - h_t$, then $\widetilde{W}_t$ is a Wiener process in the $\widetilde{P}$-measure. More generally, if X_t is the diffusion process satisfying the SDE

$$dX_t = a(X_t, t)dW_t + b(X_t, t)dt$$

under the probability measure P, then under the probability measure $\widetilde{P}$ defined in (3.255), X_t satisfies the following SDE

$$\begin{aligned} dX_t &= a(X_t, t)dW_t + b(X_t, t)dt \\ &= a(X_t, t)\left(d\widetilde{W}_t + h_t dt\right) + b(X_t, t)dt \\ &= a(X_t, t)d\widetilde{W}_t + \left(b(X_t, t) + h_t a(X_t, t)\right)dt. \end{aligned}$$

In other words, in the P-measure, X_t has drift part b and diffusion part a; whereas in the $\widetilde{P}$-measure, the diffusion part stays the same while the drift part becomes $b + h\,a$.

In particular, if we choose h to be $-\frac{b}{a}$, then X_t becomes driftless in the $\widetilde{P}$-measure provided that the stochastic exponential $\mathcal{E}_t(h)$ is a martingale.

We note that the martingality[21] of the stochastic exponential $\mathcal{E}_t(h)$ guarantees that the new measure $\widetilde{P}$ defined in (3.255) is a probability measure, i.e., $\int_\Omega d\widetilde{P} = 1$, since

$$\begin{aligned}
\int_\Omega d\widetilde{P} &= \int_\Omega \frac{d\widetilde{P}}{dP} dP \\
&= E\left[\frac{d\widetilde{P}}{dP}\right] \\
&= E\left[\mathcal{E}_T(h)\right] \\
&= \mathcal{E}_0(h) \\
&= 1. \quad \square
\end{aligned}$$

Question 26. What is the martingale representation theorem, and how is it related to option pricing and hedging?

Answer: Let W_t be a Wiener process defined on the filtered probability space $(\Omega, \mathcal{F}_t, P)$, where the filtration $\{\mathcal{F}_t\}$ is generated by the Wiener process W_t. Let M_t be a martingale with respect to $\{\mathcal{F}_t\}$ such that M_t is square integrable for every t, i.e., $E[M_t^2] < \infty$, for all t. The martingale representation theorem asserts that, for any such a martingale M_t, there exists an $\mathcal{F}_t$-adapted square integrable process θ_t such that M_t has the stochastic integral representation

$$M_t = E[M_0] + \int_0^t \theta_s dW_s$$

almost surely. Note that such martingales have to be continuous.

[21] A sufficient condition which ensures the martingality of $\mathcal{E}_t(h)$ is the Novikov's condition $E\left[e^{\frac{1}{2}\int_0^T h_t^2 dt}\right] < \infty$.

The relationship between the martingale representation theorem and option pricing and hedging is as follows. For simplicity, assume that the risk free rate is zero. Assume that the price process S_t of the underlying asset follows the diffusion process determined by the SDE

$$dS_t = \sigma_t S_t dW_t$$

under risk neutral probability P, where the driving Wiener process W_t is defined on the filtered probability space $(\Omega, \mathcal{F}_t, P)$ with the filtration $\{\mathcal{F}_t\}$ is generated by W_t. Consider a derivative whose payoff function (possibly path dependent) is φ_T at maturity T. The Fundamental Theorem of Asset Pricing asserts that, because the risk free rate is assumed zero, the value process V_t of the derivative security is a martingale under risk neutral probability. In fact,

$$V_t = E[\varphi_T|\mathcal{F}_t].$$

Moreover, if the payoff function φ_T is square integrable, i.e., $E[\varphi_T^2] < \infty$, then V_t is a square integrable martingale. Therefore, by applying the martingale representation theorem, we find that there exists a $\mathcal{F}_t$-adapted process θ_t such that

$$\begin{aligned}
\varphi_T &= V_T = E[V_0] + \int_0^T \theta_t dW_s \\
&= E[V_0] + \int_0^T \frac{\theta_t}{\sigma_t S_t} dS_t.
\end{aligned}$$

Note that the quantity $\frac{\theta_t}{\sigma_t S_t}$ indicates the amount of shares to hold in order to dynamically hedge the position of the derivative with payoff function φ_T at maturity T. In particular, if φ_T is the payoff of a call option, then $\frac{\theta_t}{\sigma_t S_t}$ corresponds to the delta of the call. □

Question 27. Solve

$$dY_t = Y_t dW_t, \tag{3.256}$$

where W_t is a Wiener process.

Answer:

Solution 1: Note that (3.256) is a particular case for $\mu = 0$ and $\sigma = 1$ of the stochastic differential equation

$$dS_t = \mu S_t dt + \sigma S_t dW_t, \tag{3.257}$$

which is the model for the evolution of an asset in the Black–Scholes framework. The solution of (3.257) has the distribution

$$S_t = S_0 \exp\left(\left(\mu - \frac{\sigma^2}{2}\right)t + \sigma\sqrt{t}Z\right), \tag{3.258}$$

with $t > 0$, where Z is the standard normal variable. For $\mu = 0$ and $\sigma = 1$ in (3.258), we obtain that the solution to (3.256) is

$$Y_t = Y_0 \exp\left(-\frac{t}{2} + \sqrt{t}Z\right). \tag{3.259}$$

Solution 2: From Itô's lemma it follows that, if Y_t is a stochastic process satisfying $dY_t = Y_t dW_t$ and $f(y)$ is a function with continuous second order derivative, then

$$d(f(Y_t)) = \frac{1}{2}f''(Y_t)Y_t^2 dt + f'(Y_t)Y_t dW_t. \tag{3.260}$$

For $f(y) = \ln(y)$, we obtain from (3.260) that

$$d(\ln(Y_t)) = -\frac{1}{2}dt + dW_t. \tag{3.261}$$

By integrating (3.261) between 0 and t, we find that

$$\ln(Y_t) - \ln(Y_0) = -\frac{t}{2} + W_t - W_0 = -\frac{t}{2} + W_t,$$

since $W_0 = 0$, and therefore

$$Y_t = Y_0 \exp\left(-\frac{t}{2} + W_t\right).$$

Since W_t is a Wiener process, W_t is a normal random variable of mean 0 and variance t, i.e., $W_t = \sqrt{t}Z$, where Z is the standard normal variable. Thus, Y_t has the distribution

$$Y_t = Y_0 \exp\left(-\frac{t}{2} + \sqrt{t}Z\right),$$

which is the same as (3.259). □

Question 28. Solve the following SDEs:

(i) $dY_t = \mu Y_t dt + \sigma Y_t dW_t$;

(ii) $dX_t = \mu dt + (aX_t + b)dW_t$.

Answer:
(i) Recall from Itô's formula that, if Y_t satisfies the SDE

$$dY_t = \mu Y_t dt + \sigma Y_t dW_t,$$

and if $f(y)$ is a function with continuous second order derivative, then

$$\begin{aligned} df(Y_t) &= f'(Y_t)dY_t + \frac{1}{2}f''(Y_t)\sigma^2 Y_t^2 dt \\ &= \left[\mu Y_t f'(Y_t) + \frac{\sigma^2}{2}Y_t^2 f''(Y_t)\right] dt \\ &+ \sigma Y_t f'(Y_t)dW_t. \end{aligned} \tag{3.262}$$

For $f(y) = \ln(y)$, we obtain from (3.262) that

$$d\ln(Y_t) = \left(\mu - \frac{\sigma^2}{2}\right) dt + \sigma dW_t. \tag{3.263}$$

By integrating (3.263) from 0 to t, we obtain that

$$\ln(Y_t) - \ln(Y_0) = \left(\mu - \frac{\sigma^2}{2}\right) t + \sigma W_t, \tag{3.264}$$

and by solving (3.264) for Y_t, we conclude that

$$Y_t = Y_0 e^{\left(\mu - \frac{\sigma^2}{2}\right)t + \sigma W_t}.$$

(ii) We look for a solution of

$$dX_t = \mu dt + (aX_t + b)dW_t \tag{3.265}$$

of the form

$$X_t = U_t V_t, \tag{3.266}$$

where the process U_t is defined by the solution to the SDE

$$dU_t = aU_t dW_t, \quad \text{with } U_0 = 1, \tag{3.267}$$

and the process V_t is the solution to the SDE

$$dV_t = \alpha_t dt + \beta_t dW_t, \quad \text{with } V_0 = X_0, \tag{3.268}$$

where the coefficients α_t and β_t are to be determined.

Recall from (i) that the solution to

$$dY_t = \mu Y_t dt + \sigma Y_t dW_t$$

is

$$Y_t = Y_0 \, e^{\left(\mu - \frac{\sigma^2}{2}\right)t + \sigma W_t}. \tag{3.269}$$

By letting $\mu = 0$ and $\sigma = a$ in (3.269), we find that the solution to (3.267) is

$$U_t = e^{aW_t - \frac{a^2}{2}t}. \tag{3.270}$$

Note that

$$\begin{aligned} dU_t \, dV_t &= (aU_t dW_t)(\alpha_t dt + \beta_t dW_t) \\ &= a\beta_t U_t dt, \end{aligned} \tag{3.271}$$

since $(dW_t)^2 = dt$ and $dW_t dt = 0$ (because this term has order $(dt)^{3/2}$). By applying Itô's product rule to $X_t = U_t V_t$ and using (3.271), we obtain that

$$\begin{aligned} dX_t &= d(U_tV_t) = U_tdV_t + V_tdU_t + dU_t\,dV_t \\ &= U_t(\alpha_t dt + \beta_t dW_t) + V_t(aU_t dW_t) \\ &\quad + a\beta_t U_t dt \\ &= (a\beta_t + \alpha_t)U_t dt + (aU_tV_t + \beta_t U_t)dW_t \\ &= (a\beta_t + \alpha_t)U_t dt + (aX_t + \beta_t U_t)dW_t \quad (3.272) \end{aligned}$$

Then, $X_t = U_t V_t$ is a solution to (3.265) if and only if the coefficients of the dW_t terms and of the dt terms in (3.265) and (3.272) are equal.

From the dW_t terms, we obtain that

$$\begin{aligned} & aX_t + b = aX_t + \beta_t U_t \\ \Longleftrightarrow\quad & \beta_t U_t = b \\ \Longleftrightarrow\quad & \beta_t = bU_t^{-1} = b\,e^{-aW_t + \frac{a^2}{2}t}. \end{aligned} \qquad (3.273)$$

From the dt terms, we obtain that

$$\begin{aligned} & (a\beta_t + \alpha_t)U_t = \mu \\ \Longleftrightarrow\quad & a\beta_t + \alpha_t = \mu U_t^{-1} \\ \Longleftrightarrow\quad & \alpha_t = \mu\, U_t^{-1} - a\beta_t \\ \Longleftrightarrow\quad & \alpha_t = \mu\, e^{-aW_t + \frac{a^2}{2}t} - abe^{-aW_t + \frac{a^2}{2}t} \\ \Longleftrightarrow\quad & \alpha_t = (\mu - ab)\, e^{-aW_t + \frac{a^2}{2}t}. \end{aligned} \qquad (3.274)$$

Note that the solution V_t to (3.268) is given by

$$V_t = X_0 + \int_0^t \alpha_s ds + \int_0^t \beta_s dW_s. \qquad (3.275)$$

From (3.266), (3.270), and (3.275), we find that the solution X_t to (3.265) is

$$X_t = e^{aW_t - \frac{a^2}{2}t}\left(X_0 + \int_0^t \alpha_s ds + \int_0^t \beta_s dW_s\right),$$

and, using (3.273) and (3.274), we obtain that

$$\begin{aligned} X_t &= e^{aW_t - \frac{a^2}{2}t}\left[X_0 + \int_0^t (\mu - ab)\, e^{-aW_s + \frac{a^2}{2}s} ds \right. \\ &\quad \left. + \int_0^t be^{-W_s + \frac{a^2}{2}s} dW_s\right] \\ &= X_0\, e^{aW_t - \frac{a^2}{2}t} \\ &\quad + (\mu - ab) \int_0^t e^{a(W_t - W_s) - \frac{a^2}{2}(t-s)} ds \\ &\quad + b \int_0^t e^{a(W_t - W_s) - \frac{a^2}{2}(t-s)} dW_s. \quad \square \end{aligned}$$

Question 29. What is the Heston model?

Answer: In Heston's stochastic volatility model, it is assumed that the price of the underlying asset satisfies the same SDE as in the lognormal model, i.e.,

$$dS_t = \mu S_t dt + \sqrt{v_t} S_t dW_t,$$

whereas the instantaneous variance v_t itself follows a mean reverting CIR (Cox-Ingersoll-Ross) process

$$dv_t = \lambda(v_t - m)dt + \eta\sqrt{v_t} dZ_t,$$

where $\lambda > 0$ and $m > 0$ are positive constants. The driving Wiener processes W_t and Z_t are correlated with constant correlation ρ, i.e.,

$$\mathrm{corr}(dW_t, dZ_t) = \rho dt.$$

The Heston model is a benchmark, and is commonly used in derivative pricing because it has following features:

- it takes into account the leverage effect, namely, the driving Wiener processes W_t and Z_t are correlated; empirically, the correlation ρ is negative and that is why the word "leverage";

- it has a quasi closed form (up to an inverse Fourier transform) solution for the prices of European options, which make the calibration more tractable;

- the variance process is mean reverting with rate of reversion λ and long term mean m.

Note that, if the parameters of the volatility process are in the regime $2\lambda m < \eta^2$, then zero is an attainable boundary for the volatility process. Practitioners usually assume that the boundary behavior at zero is either absorption, i.e., the process is stuck at zero once it hits zero, or reflection, i.e, the process bounces back right after it hits zero. On the other hand, the other boundary infinity is an unattainable boundary. □

Question 30. Show that the probability density function of the standard normal integrates to 1.

Answer: The probability density function of the standard normal variable is $\frac{1}{\sqrt{2\pi}}\, e^{-\frac{t^2}{2}}$. We want to show that

$$\frac{1}{\sqrt{2\pi}} \int_{-\infty}^{\infty} e^{-\frac{t^2}{2}}\, dt = 1,$$

which, using the substitution $t = \sqrt{2}x$, can be written as

$$I = \int_{-\infty}^{\infty} e^{-x^2}\, dx = \sqrt{\pi}. \qquad (3.276)$$

We prove (3.276) by using polar coordinates. Since x is just an integrating variable, we can also write the integral I in terms of another integrating variable, denoted by y, as $I = \int_{-\infty}^{\infty} e^{-y^2}\, dy$.

Then,[22]

$$
\begin{aligned}
I^2 &= \left(\int_{-\infty}^{\infty} e^{-x^2}\,dx\right)\cdot\left(\int_{-\infty}^{\infty} e^{-y^2}\,dy\right) \\
&= \int_{-\infty}^{\infty}\int_{-\infty}^{\infty} e^{-x^2}e^{-y^2}\,dxdy \qquad (3.277) \\
&= \int_{-\infty}^{\infty}\int_{-\infty}^{\infty} e^{-(x^2+y^2)}\,dxdy.
\end{aligned}
$$

We use the polar coordinates transformation $x = r\cos\theta$ and $y = r\sin\theta$, with $r \in [0, \infty)$ and $\theta \in [0, 2\pi)$, to evaluate the last integral. Since $dxdy = rd\theta dr$, we obtain that

$$
\begin{aligned}
I^2 &= \int_{-\infty}^{\infty}\int_{-\infty}^{\infty} e^{-(x^2+y^2)}\,dxdy \\
&= \int_0^{\infty}\int_0^{2\pi} r\,e^{-\left(r^2\cos^2\theta+r^2\sin^2\theta\right)}\,d\theta dr \\
&= \int_0^{\infty}\int_0^{2\pi} r\,e^{-r^2}\,d\theta dr \qquad (3.278) \\
&= \int_0^{\infty} 2\pi\,r\,e^{-r^2}\,dr \\
&= 2\pi\lim_{t\to\infty}\int_0^t r\,e^{-r^2}\,dr \\
&= 2\pi\lim_{t\to\infty}\left(-\frac{1}{2}e^{-r^2}\right)\Bigg|_0^t \\
&= \pi;
\end{aligned}
$$

note that (3.278) follows from the equality $\cos^2\theta+\sin^2\theta = 1$ for any real number θ.

Since $I > 0$, $I = \sqrt{\pi}$, which is what we wanted to prove; see (3.276). □

[22] Note that Fubini's theorem is needed for a rigorous derivation of the equality (3.277); this technical step is rarely required by the interviewer.

Question 31. Let $W_t = (X_t, Y_t)$ be a two dimensional Brownian motion starting at (x, y), i.e., X_t and Y_t are independent one dimensional Brownian motions with $X_0 = x$ and $Y_0 = y$.

(i) Find the probability that the Brownian motion W_t reaches the y-axis before reaching the x-axis.

(ii) Let $0 < r_1 < r_2$ such that $r_1 < \sqrt{x^2 + y^2} < r_2$. Find the probability that W_t enters the inner circle of center 0 and radius r_1 before leaving the outer circle of center 0 and radius r_2.

Answer: (i) Denote by τ_x and τ_y the stopping times

$$\begin{aligned} \tau_x &= \inf\{t > 0 : Y_t = 0\}; \\ \tau_y &= \inf\{t > 0 : X_t = 0\} \end{aligned}$$

and let

$$\tau = \min\{\tau_x, \tau_y\}.$$

In other words, τ_x is the first hitting time of the x–axis for the Brownian motion W_t, τ_y is the first hitting time of the y–axis for W_t, and τ is the first time W_t hits either the x–axis or the y–axis.

We are asked to find the probability

$$\mathbb{P}_{(x,y)}[\tau_y < \tau_x]$$

where $\mathbb{P}_{(x,y)}[\cdot]$ denotes the (conditional) probability corresponding to the two dimensional Brownian motion W_t starting at the point (x, y) at time 0.

Note that

$$\mathbb{P}_{(x,y)}[\tau_y < \tau_x] + \mathbb{P}_{(x,y)}[\tau_y > \tau_x] = 1$$

since a two dimensional Brownian motion is recurrent and therefore the two dimensional Brownian motion W_t must hit either the x-axis or the y-axis in finite time almost surely.

Let $u(x, y)$ be a function with continuous and bounded second order derivatives. By applying Itô's formula to $u(X_t, Y_t)$ with stopping time τ, we obtain that

$$u(X_\tau, Y_\tau) - u(x, y) \tag{3.279}$$

$$= \int_0^\tau u_x dX_t + \int_0^\tau u_y dY_t \tag{3.280}$$

$$+\frac{1}{2}\int_0^\tau \{u_{xx} + u_{yy}\}\, dt. \tag{3.281}$$

Assume that the function $u(x, y)$ satisfies the PDE $u_{xx} + u_{yy} = 0$ in the first quadrant of the (x, y) plane with boundary conditions $u(0, y) = 1$ and $u(x, 0) = 0$. By taking the conditional expectation $\mathbb{E}_{(x,y)}[\cdot]$ on both sides of (3.279–3.281) and rearranging terms of the resulting equation, we obtain that

$$\begin{aligned} u(x, y) &= \mathbb{E}_{(x,y)}\left[u(X_\tau, Y_\tau)\right] \\ &= 1 \cdot \mathbb{P}_{(x,y)}\left[\tau = \tau_y\right] + 0 \cdot \mathbb{P}_{(x,y)}\left[\tau = \tau_x\right] \\ &= \mathbb{P}_{(x,y)}[\tau_y < \tau_x]. \end{aligned}$$

To solve the boundary value problem for u, we transform the equation into polar coordinates $u = u(r, \theta)$ as

$$u_{rr} + \frac{1}{r}u_r + \frac{1}{r^2}u_{\theta\theta} = 0$$

and the boundary conditions become $u(r, \pi/2) = 1$ and $u(r, 0) = 0$, respectively. Notice that since the solution u is radially symmetric, we may use the ansatz $u(r, \theta) = f(\theta)$, for some function f. Thus, the problem reduces to the following second order linear ODE

$$f''(\theta) = 0$$

with boundary conditions $f(\pi/2) = 1$ and $f(0) = 0$. The solution is given by $f(\theta) = \frac{2}{\pi}\theta$. Thus,

$$u(x, y) = \frac{2}{\pi}\arctan\left(\frac{y}{x}\right).$$

(ii) Denote by τ_1 and τ_2 the stopping times

$$\begin{aligned}\tau_1 &= \inf\{t > 0 : |W_t| < r_1\};\\ \tau_2 &= \inf\{t > 0 : |W_t| > r_2\},\end{aligned}$$

and let

$$\tau = \min\{\tau_1, \tau_2\}.$$

We are asked to find the probability

$$\mathbb{P}_{(x,y)}[\tau_1 < \tau_2].$$

By the same token as in part (i) of the problem, the probability is determined by solving the PDE in polar coordinates

$$u_{rr} + \frac{1}{r}u_r + \frac{1}{r^2}u_{\theta\theta} = 0$$

with boundary conditions $u(r_1, \theta) = 1$ and $u(r_2, \theta) = 0$ for all θ. Since the solution u is rotationally symmetric, we use the ansatz $u(r, \theta) = g(r)$. The problem reduces to the following second order linear ODE

$$g''(r) + \frac{1}{r}g'(r) = 0, \quad \forall\, r_1 < r < r_2,$$

with boundary conditions

$$g(r_1) = 1; \quad g(r_2) = 0.$$

The solution is given by

$$g(r) = \frac{\ln r - \ln r_2}{\ln r_1 - \ln r_2}, \quad \forall\, r_1 < r < r_2.$$

Consequently,

$$u(x, y) = \frac{\ln\sqrt{x^2 + y^2} - \ln r_2}{\ln r_1 - \ln r_2}. \quad \square$$

Question 32. Let B_t, $t \geq 0$, be a standard Brownian motion in the probability measure $\mathbb{P}$. Determine the probability density function of $B_{\frac{3}{2}}$ under the probability measure $\widetilde{\mathbb{P}}$ defined by the Radon–Nikodym derivative

$$\frac{d\widetilde{\mathbb{P}}}{d\mathbb{P}} = e^{B_1 - \frac{1}{2}}. \tag{3.282}$$

Answer: Let

$$\theta_t = 1_{[0,1]}(t), \; t \in \left[0, \frac{3}{2}\right].$$

Then,

$$B_1 - \frac{1}{2} = \int_0^{\frac{3}{2}} \theta_t dB_t - \frac{1}{2} \int_0^{\frac{3}{2}} \theta_t^2 dt$$

and therefore the Radon-Nikodym derivative from (3.282) can be written as

$$\frac{d\widetilde{\mathbb{P}}}{d\mathbb{P}} = \exp\left(\int_0^{\frac{3}{2}} \theta_t dB_t - \frac{1}{2} \int_0^{\frac{3}{2}} \theta_t^2 dt \right),$$

which defines a probability measure.

From Girsanov's theorem, it follows that

$$\widetilde{B}_t = B_t + \int_0^t \theta_s ds \tag{3.283}$$

and $\widetilde{B}_t$ is a standard Brownian motion under $\widetilde{\mathbb{P}}$.

From (3.283), we obtain that

$$B_{\frac{3}{2}} = \widetilde{B}_{\frac{3}{2}} - 1,$$

and we conclude that $B_{\frac{3}{2}}$ is normally distributed with mean -1 and variance $\frac{3}{2}$ under $\widetilde{\mathbb{P}}$.

Intuitively, one can think of the result as follows:

• Since the change of measure only changes the drift and does not affect the diffusion part, $B_{\frac{3}{2}}$ must a Gaussian random variable of variance $\frac{3}{2}$.

• The change of measure is in effect only before time 1, when B_t is a Brownian motion with drift -1 under $\widetilde{\mathbb{P}}$, and after time 1 it switches back to a standard Brownian motion starting at -1. □

Question 33. Let $\boldsymbol{B}_t = (B_t^{(1)}, B_t^{(2)})$ be a two dimensional Brownian motion in the (x, y) plane. Let $a > 0$ and denote by τ the first time $\boldsymbol{B}_t$ hits the line $y = a$. Determine the probability distribution of $B_\tau^{(1)}$.

Answer: Since the two Brownian motions $B_t^{(1)}$ and $B_t^{(2)}$ are independent, it follows that the Brownian motion $B_\tau^{(1)}$ is independent of the random time τ. Hence, we can calculate the probability distribution of $B_\tau^{(1)}$ by conditioning on τ as follows:

$$\begin{aligned}
\mathbb{P}\left[B_\tau^{(1)} \leq x\right] &= \int_0^\infty \mathbb{P}\left[B_\tau^{(1)} \leq x | \tau = t\right] f_\tau(t)\, dt \\
&= \int_0^\infty \mathbb{P}\left[B_t^{(1)} \leq x\right] f_\tau(t)\, dt \\
&= \int_0^\infty \Phi\left(\frac{x}{\sqrt{t}}\right) f_\tau(t) dt, \qquad (3.284)
\end{aligned}$$

where f_τ is the probability density function of τ and Φ is the cumulative distribution function of the standard normal distribution.

Denote by $f_{B_\tau^{(1)}}$ the probability density function of

$B_\tau^{(1)}$. Then, from (3.284), it follows that

$$\begin{aligned} f_{B_\tau^{(1)}}(x) &= \frac{d}{dx}\mathbb{P}\left[B_\tau^{(1)} \le x\right] \\ &= \frac{d}{dx}\left(\int_0^\infty \Phi\left(\frac{x}{\sqrt{t}}\right) f_\tau(t)dt\right) \\ &= \int_0^\infty \frac{1}{\sqrt{t}}\phi\left(\frac{x}{\sqrt{t}}\right) f_\tau(t)dt, \end{aligned} \qquad (3.285)$$

where ϕ denotes the probability density function of the standard normal distribution.

To determine the probability density function f_τ of τ, note the equivalence between the following two events:

$$\{\tau \le t\} = \left\{\max_{0\le s\le t} B_s^{(2)} \ge a\right\}. \qquad (3.286)$$

From the reflection principle for the Brownian motion, it follows that

$$\mathbb{P}\left[\max_{0\le s\le t} B_s^{(2)} \ge a\right] = 2\mathbb{P}\left[B_t^{(2)} \ge a\right], \qquad (3.287)$$

and therefore, from (3.286) and (3.287), we find that

$$\begin{aligned} \mathbb{P}[\tau \le t] &= \mathbb{P}\left[\max_{0\le s\le t} B_s^{(2)} \ge a\right] \\ &= 2\,\mathbb{P}\left[B_t^{(2)} \ge a\right] \\ &= 2\left(1 - \mathbb{P}\left[B_t^{(2)} \le a\right]\right) \\ &= 2\left(1 - \Phi\left(\frac{a}{\sqrt{t}}\right)\right). \end{aligned}$$

Thus, the probability density function of τ is given by

$$\begin{aligned} f_\tau(t) &= \frac{d}{dt}\mathbb{P}[\tau \le t] \\ &= \frac{d}{dt}\left[2\left(1 - \Phi\left(\frac{a}{\sqrt{t}}\right)\right)\right] \\ &= \frac{a}{t^{3/2}}\,\phi\left(\frac{a}{\sqrt{t}}\right). \end{aligned} \qquad (3.288)$$

From (3.285) and (3.288), we obtain that

$$\begin{aligned} f_{B_\tau^{(1)}}(x) &= \int_0^\infty \frac{1}{\sqrt{t}}\phi\left(\frac{x}{\sqrt{t}}\right) f_\tau(t)dt \\ &= \int_0^\infty \frac{1}{\sqrt{t}}\phi\left(\frac{x}{\sqrt{t}}\right) \frac{a}{t^{3/2}}\,\phi\left(\frac{a}{\sqrt{t}}\right) dt \\ &= \frac{a}{2\pi}\int_0^\infty \frac{1}{t^2} e^{-\frac{x^2+a^2}{2t}}\, dt \\ &= \frac{a}{2\pi}\cdot\frac{2}{x^2+a^2}\left(e^{-\frac{x^2+a^2}{2t}}\Big|_{t=0}^{t=\infty}\right) \\ &= \frac{a}{\pi(x^2+a^2)}. \end{aligned}$$

We conclude that $B_\tau^{(1)}$ has a Cauchy distribution with location parameter 0 and scale parameter a. □

Question 34. If Z is a standard normal variable, find the covariance of Z and Z^2.

Answer: The covariance of two random variables X and Y is

$$\begin{aligned} \mathrm{cov}(X,Y) &= E[(X-E[X])(Y-E[Y])] \\ &= E[XY] - E[X]\,E[Y]. \end{aligned}$$

Thus,

$$\mathrm{cov}(Z,Z^2) = E[Z^3] - E[Z]\,E[Z^2] = 0,$$

since $E[Z] = 0$ and $E[Z^3] = 0$. □

Question 35. If the function

$$f(x) = \frac{a}{1+x^2},$$

where a is a real constant, is the probability density function of a random variable, find the value of a.

Answer: If $f(x) = \frac{a}{1+x^2}$ is the probability density function of a random variable, then the integral of $f(x)$ over the entire real axis is equal to 1, i.e.,

$$\int_{-\infty}^{\infty} f(x)\,dx = a\int_{-\infty}^{\infty} \frac{1}{1+x^2}\,dx = 1. \qquad (3.289)$$

Recall that

$$(\arctan(x))' = \frac{1}{1+x^2}.$$

Thus,

$$\begin{aligned} \int_{-\infty}^{\infty} \frac{1}{1+x^2} &= \arctan(x)\,\Big|_{-\infty}^{\infty} \\ &= \frac{\pi}{2} - \left(-\frac{\pi}{2}\right) \\ &= \pi. \end{aligned} \qquad (3.290)$$

From (3.289) and (3.290), we obtain that $a \cdot \pi = 1$ and therefore

$$a = \frac{1}{\pi}. \quad \square$$

Question 36. How many independent random variables uniformly distributed on $[0,1]$ should you generate to ensure that there is at least one between 0.70 and 0.72 with probability 95%?

Answer: Denote by N the smallest number of random variables you should generate such that

$$P(\text{at least one r.v. in } [0.70, 0.72]) \geq 0.95. \qquad (3.291)$$

The probability that a random variable uniformly distributed on $[0,1]$ is not in the interval $[0.70, 0.72]$ is 0.98. Thus, the probability that none of the N independent variables are in $[0.70, 0.72]$ is 0.98^N, i.e.,

$$P(\text{no r.v. in } [0.70, 0.72]) = 0.98^N.$$

Note that

$$\begin{aligned} & P(\text{at least one r.v. in } [0.70, 0.72]) \\ = \ & 1 - P(\text{no r.v. in } [0.70, 0.72]) \\ = \ & 1 - (0.98)^N. \end{aligned} \tag{3.292}$$

From (3.291) and (3.292), we find that N is the smallest integer such that

$$1 - (0.98)^N \ \geq \ 0.95,$$

which is equivalent to

$$\begin{aligned} & (0.98)^N \leq 0.05 \\ \Longleftrightarrow \quad & N \ln(0.98) \leq \ln(0.05) \\ \Longleftrightarrow \quad & N \geq \frac{\ln(0.05)}{\ln(0.98)} \approx 148.28 \\ \Longleftrightarrow \quad & N = 149. \end{aligned}$$

We conclude that at least 149 uniform random variables on $[0, 1]$ must be generated in order to have 95% confidence that at least one of the random variables is between 0.70 and 0.72. □

3.8 Brainteasers.

Question 1. A flea is going between two points which are 100 inches apart by jumping (always in the same direction) either one inch or two inches at a time. How many different paths can the flea travel by?

Answer: Let a_n denote the number of different paths of the flea that covers the distance of n inches, jumping either one inch or two inches at a time. We want to find a_{100}.

Since the flea can jump either one inch or two inches at a time, it could have made the last jump either from the end of the $(n-1)$st inch or from the end of the $(n-2)$nd inch. Hence, the total number of ways the flea can cover the distance of n inches, jumping either one inch or two inches at a time, is the sum of the number of ways the flea can cover the distance of $n-1$ inches, jumping either one inch or two inches at a time, and the number of ways the flea can cover the distance of $n-2$ inches, jumping either one inch or two inches at a time.

In other words, for $n > 2$, we have

$$a_n = a_{n-1} + a_{n-2}. \tag{3.293}$$

Note that $a_1 = 1$, since the flea can cover 1 inch in only one way, jumping one inch once; while $a_2 = 2$, since the flea can cover 2 inches in two ways, either jumping one inch twice, or jumping two inches once.

Note that (3.293) is the recurrence relation for the Fibonacci sequence. Let

$$\phi_1 = \frac{1+\sqrt{5}}{2} \quad \text{and} \quad \phi_2 = \frac{1-\sqrt{5}}{2}$$

be the roots of the characteristic equation $x^2 - x - 1 = 0$ corresponding to (3.293).

Then,

$$a_n = C_1\phi_1^n + C_2\phi_2^n, \ \forall\, n \geq 1,$$

where the constants C_1 and C_2 are such that $a_1 = 1$ and $a_2 = 2$.

By solving the linear system

$$\left\{ \begin{array}{lcl} C_1\phi_1 + C_2\phi_2 & = & 1 \\ C_1\phi_1^2 + C_2\phi_2^2 & = & 2, \end{array} \right.$$

we obtain that

$$\begin{aligned} C_1 &= \frac{1+\sqrt{5}}{2\sqrt{5}} = \frac{\phi_1}{\sqrt{5}}; \\ C_2 &= -\frac{\sqrt{5}-1}{2\sqrt{5}} = -\frac{\phi_2}{\sqrt{5}}, \end{aligned}$$

and therefore

$$a_n = \frac{1}{\sqrt{5}}\left(\phi_1^{n+1} - \phi_2^{n+1}\right).$$

Plugging $n = 100$ into the last expression gives the answer. □

Question 2. I have a bag containing three pancakes: one golden on both sides, one burnt on both sides, and one golden on one side and burnt on the other. You shake the bag, draw a pancake at random, look at one side, and notice that it is golden. What is the probability that the other side is golden?

Answer:

Solution 1: Label the pancakes 0, 1, and 2, according to the number of burnt sides. Let E_i, $i = 0, 1, 2$, denote the event that the pancake with i burnt sides is drawn from the bag. Let A denote the event that the side (of the randomly drawn pancake) we look at is golden.

The probability of the pancake drawn from the bag having no burnt sides given that one side is golden is $P(E_0|A)$. Then, from the Bayes'formula, we obtain that

$$
\begin{aligned}
P(E_0|A) &= \frac{P(E_0 \cap A)}{P(A)} = \\
&\frac{P(E_0)P(A|E_0)}{P(A|E_0)P(E_0) + P(A|E_1)P(E_1) + P(A|E_2)P(E_2)} \\
&= \frac{\frac{1}{3}\cdot 1}{1\cdot\frac{1}{3} + \frac{1}{2}\cdot\frac{1}{3} + 0\cdot\frac{1}{3}} \\
&= \frac{2}{3}.
\end{aligned}
$$

Solution 2: Out of the six possible sides that we could have seen, three are golden. Out of these, two belong to a pancake that is golden on both sides. Therefore, the probability of the other side being golden is $\frac{2}{3}$. □

Question 3. Alice and Bob are playing heads and tails, Alice tosses $n+1$ coins, Bob tosses n coins. The coins are fair. What is the probability that Alice will have strictly more heads than Bob?

Answer:

Solution 1: Alice flips the coin more often than Bob, so either she must end up with more heads or with more tails than Bob. She cannot, however, end up with more heads and more tails, because she only flips one more coin than Bob. We deduce that either Alice gets more heads or she gets more tails.

Since these events are equally likely, they both have probability $\frac{1}{2}$, and therefore the probability that Alice will have strictly more heads than Bob is $\frac{1}{2}$.

Solution 2: Suppose that Alice and Bob begin by flipping n coins each. Let p be the probability that Alice gets more heads than Bob, and let q be the probability that both Alice and Bob get an equal number of heads. Note that $2p + q = 1$ since the probability that Alice gets more heads than Bob is, by symmetry, equal to the probability that Bob gets more heads than Alice.

Alice then flips the $(n+1)$st coin. For Alice to have more heads, she either had more heads than Bob before flipping the last coin; or the same number of heads as Bob before flipping the last coin and her $(n+1)$st flip must come up heads. Hence, the probability of Alice winning is $p + \frac{1}{2}q$.

Since $2p + q = 1$, then $p + \frac{1}{2}q = \frac{1}{2}$. $\square$

Question 4. Alice is in a restaurant trying to decide between three desserts. How can she choose one of three desserts with equal probability with the help of a fair coin? What if the coin is biased and the bias is unknown?

Answer:
Solution 1: Denote the desserts by A, B, and C. First, suppose the coin is fair. Denote heads by H and tails by T. The procedure to choose one of three desserts with equal probability is as follows: toss the coin twice; let the outcomes TH, HT, and TT correspond to choosing desserts A, B, and C, respectively; if the outcome is HH, repeat the procedure.

Note that the probability of our procedure not being repeated is $p = \frac{3}{4}$; hence, the number of times our procedure is repeated is a geometric random variable with p as its parameter. The expected number of times our procedure is repeated is $\frac{1}{p} = \frac{4}{3}$. Since each procedure involves two coin tosses, the expected number of coin tosses before Alice chooses one of three desserts with equal probability is $\frac{8}{3}$.

Now, suppose the coin is biased. One procedure to choose one of three desserts with equal probability would be as follows: toss the coin four times; denote by $THHT$, $HTTH$, and $THTH$ the outcomes corresponding to choosing desserts A, B, and C, respectively; all the other 4-toss outcomes result in repeating the procedure.

Solution 2: An alternative procedure is as follows: toss the coin three times; denote by HTT, THT, and TTH the outcomes corresponding to choosing desserts A, B, and C, respectively; all the other 3-toss outcomes result in repeating the procedure.

Using an argument similar to the case with a fair coin, one finds that the expected number of tosses of a coin with an unknown bias, before Alice chooses one of three desserts with equal probability, is $\frac{64}{3}$ and 8, respectively, for the two procedures described above. □

Question 5. What is the expected number of times you must flip a fair coin until it lands on head? What if the coin is biased and lands on head with probability p?

Answer: Denote by X be the number of times you must flip a fair coin until it lands on head. If the first coin toss is a head (which happens with probability $\frac{1}{2}$), then $X = 1$. If the first coin toss is a tail (which also happens with probability $\frac{1}{2}$), then the coin tossing process resets and the number of steps before the coin lands on head will be 1 plus the expected number of coin tosses until the coin lands heads. In other words, the expected number of coin toss $E[X]$ satisfies the equation

$$E[X] \;=\; \frac{1}{2} \;+\; \frac{1}{2}\;(1 + E[X]). \tag{3.294}$$

Solving (3.294) for $E[X]$, we conclude that $E[X] = 2$, i.e., the expected number of times you must flip a fair coin until it lands heads is 2.

On the other hand, if the coin is biased with the probability p of landing on head, the same argument still applies. However, in this case, (3.294) reads

$$E[X] \;=\; p \;+\; (1 - p)\;(1 + E[X]).$$

Again, solve for $E[X]$, we obtain $E[X] = \frac{1}{p}$. □

Question 6. What is the expected number of coin tosses of a fair coin in order to get two heads in a row? What if the coin is biased with 25% probability of getting heads?

Answer: We solve the general case of a biased coin with probability p of the coin toss resulting in heads. The outcomes of the first two tosses are as follows:

- If the first toss is tails, which happens with probability $1-p$, then the process resets and the expected number of tosses increases by 1.
- If the first toss is heads, and if the second toss is also heads, which happens with probability p^2, then two consecutive heads are obtained after two tosses.
- If the first toss is heads, and if the second toss is tails, which happens with probability $p(1-p)$, then the process resets and the expected number of tosses increases by 2.

If $E[X]$ denotes the expected number of tosses in order to get two heads in a row, we conclude that

$$E[X] = (1-p)(1+E[X]) + 2p^2 + p(1-p)(2+E[X]).$$

By solving for $E[X]$, we obtain that

$$E[X] = \frac{1+p}{p^2}. \tag{3.295}$$

For an unbiased coin, i.e., for $p = \frac{1}{2}$, we find from (3.295) that $E[X] = 6$, i.e., the expected number of coin tosses to obtain two heads in a row is 6.

For a biased coin with 25% probability of getting heads, i.e., for $p = \frac{1}{4}$, we find from (3.295) that $E[X] = 20$, i.e., the expected number of coin tosses to obtain two heads in a row in this case is 20. □

Question 7. A fair coin is tossed n times. What is the probability that no two consecutive heads appear?

Answer: The total number of sequences of heads and tails of length n is 2^n. Let a_n be the number of sequences of heads and tails of length n, such that no two consecutive heads appear. Then, the probability that no two consecutive heads appear is $\frac{a_n}{2^n}$.

Note that $a_1 = 2$ (H and T do not contain two consecutive heads) and $a_2 = 3$ (out of HH, HT, TH, and TT, only HH contains two consecutive heads). We find the closed formula for a_n by deriving a recurrence relation as follows:

A sequence of $n \geq 3$ coin tosses does not contain two consecutive heads if and only if: (i) either it begins with a tail, followed by a sequence of $n-1$ coin tosses with no two consecutive heads; (ii) or it begins with a head, followed by a tail, and followed by a sequence of $n-2$ coin tosses with no two consecutive heads. Since these two scenarios are mutually exclusive, it follows that

$$a_n = a_{n-1} + a_{n-2}, \quad \forall\, n \geq 3. \tag{3.296}$$

Note that (3.296) is the recurrence relation for the Fibonacci sequence. Let $\phi_1 = \frac{1+\sqrt{5}}{2}$ and $\phi_2 = \frac{1-\sqrt{5}}{2}$ be the roots of the characteristic equation $x^2 - x - 1 = 0$ corresponding to (3.296). Then,

$$a_n = C_1\phi_1^n + C_2\phi_2^n, \quad \forall\, n \geq 1,$$

where the constants C_1 and C_2 are such that $a_1 = 2$ and $a_2 = 3$.

By solving the linear system

$$\begin{cases} C_1\phi_1 + C_2\phi_2 = 2 \\ C_1\phi_1^2 + C_2\phi_2^2 = 3, \end{cases}$$

we obtain that

$$C_1 = \frac{3+\sqrt{5}}{2\sqrt{5}} = \frac{\phi_1^2}{\sqrt{5}};$$

$$C_2 = -\frac{3-\sqrt{5}}{2\sqrt{5}} = -\frac{\phi_2^2}{\sqrt{5}}.$$

We conclude that

$$a_n = \frac{1}{\sqrt{5}}\left(\phi_1^{n+2} - \phi_2^{n+2}\right),$$

and therefore the probability that no two consecutive heads appear in n tosses of a fair coin, which is equal to $\frac{a_n}{2^n}$, is

$$\frac{1}{2^n\sqrt{5}}\left(\phi_1^{n+2} - \phi_2^{n+2}\right). \quad \square$$

Question 8. You have two identical Fabergé eggs, either of which would break if dropped from the top of a building with 100 floors. Your task is to determine the highest floor from which an egg could be dropped without breaking. What is the minimum number of drops required to achieve this? You are allowed to break both eggs in the process.

Answer: Consider the following more general problem:

Find the largest number of floors $h_e(n)$ a building could have in order to be able to determine the highest floor from which an egg could be dropped without breaking using e eggs and n drops.

Since one drop can only determine one floor, it follows that

$$h_e(1) = 1. \tag{3.297}$$

If we have only one egg at our disposal, the only possible strategy is to try the floors one by one from bottom to top; hence,

$$h_1(n) = n.$$

When $e \geq 2$ and $n \geq 2$, the first drop cannot be from the floor higher than $h_{e-1}(n-1) + 1$, since if the egg breaks, there are only $e - 1$ eggs and $n - 1$ drops left, and the highest floor we can still handle is $h_{e-1}(n-1)$. If the first drop does not break an egg, we can treat floor

$h_{e-1}(n-1)+2$ as the new floor 1, and reduce it to a problem with e eggs and $n-1$ drops, and therefore

$$h_e(n) \;=\; 1 + h_{e-1}(n-1) + h_e(n-1).$$

Iterating this argument, we obtain that

$$\begin{aligned} & h_e(n) \\ = \;& 1 + h_{e-1}(n-1) + h_e(n-1) \\ = \;& 2 + h_{e-1}(n-1) + h_{e-1}(n-2) + h_e(n-2) \\ = \;& \ldots \\ = \;& (n-1) + \sum_{j=1}^{n-1} h_{e-1}(j) + h_e(1) \\ = \;& n + \sum_{j=1}^{n-1} h_{e-1}(j), \end{aligned}$$

since $h_e(1) = 1$; see (3.297).

For $e = 2$, the formula above becomes

$$\begin{aligned} & h_2(n) \\ = \;& n + \sum_{j=1}^{n-1} h_1(j) = n + \sum_{j=1}^{n-1} j = n + \frac{(n-1)n}{2} \\ = \;& \frac{n(n+1)}{2}, \end{aligned} \tag{3.298}$$

where the next-to-last step follows from the summation formula $\sum_{j=1}^{k} = \frac{k(k+1)}{2}$.

Since $h_2(13) = 91 < 100 \leq 105 = h_2(14)$, the required number of drops is 14.

Note that our iterative argument also provides an algorithm: you drop the first egg from the floors

$$
\begin{aligned}
14 &= 1 + h_1(13), \\
27 &= 2 + h_1(13) + h_1(12), \\
39 &= 3 + \sum_{j=11}^{13} h_1(j), \\
50 &= 4 + \sum_{j=10}^{13} h_1(j), \\
60 &= 5 + \sum_{j=9}^{13} h_1(j), \\
69 &= 6 + \sum_{j=8}^{13} h_1(j), \\
77 &= 7 + \sum_{j=7}^{13} h_1(j), \\
84 &= 8 + \sum_{j=6}^{13} h_1(j), \\
90 &= 9 + \sum_{j=5}^{13} h_1(j), \\
95 &= 10 + \sum_{j=4}^{13} h_1(j), \\
99 &= 11 + \sum_{j=3}^{13} h_1(j),
\end{aligned}
$$

and 100; that is, move up by $14 = 1 + h_1(13)$ floors, then by $13 = 1 + h_1(12)$ floors, then by $12 = 1 + h_1(11)$ floors, and so on, until the first egg breaks (or does not) from the 100th floor. Calling the floor from which the first egg breaks f and the previously tested floor f', you drop the second egg from the intervening floors $f'+1, f'+2, \ldots, f-1$ in that order. $\square$

Question 9. An ant is in the corner of a $10 \times 10 \times 10$ room and wants to go to the opposite corner. What is the length of the shortest path the ant can take?

Answer: For clarity, assume that the ant is in a corner by the ceiling. Denote that corner by A, and denote by B the opposite corner to A, which is by the floor. The shortest path from A to B would require the ant to go on a straight line across a wall of the room to a side of the floor and from there on a straight line along the floor to the B. If you imagine laying down to the floor the vertical wall the ant went down from A to the floor, you have a 10×20 rectangle with A and B opposite corners in the rectangle. The shortest path for the ant to go from A to B is by following the diagonal of the rectangle, which has length $10\sqrt{5} \approx 22.36$. $\square$

Question 10. A $10 \times 10 \times 10$ cube is made of $1,000$ unit cubes. How many unit cubes can you see on the outside?

Answer: If all the outside unit cubes are removed, what remains is an $8 \times 8 \times 8$ cube, which is made of $8^3 = 512$ unit cubes. Thus, there are $1000 - 512 = 488$ outside unit cubes. $\square$

Question 11. Fox Mulder is imprisoned by aliens in a large circular field surrounded by a fence. Outside the fence is a vicious alien that can run four times as fast as Mulder, but is constrained to stay near the fence. If Mulder can contrive to get to an unguarded point on the fence, he can quickly scale the fence and escape. Can he get to a point on the fence ahead of the alien?

Answer: Let R denote the radius of the circular field, whose center we denote by C. Denote Mulder's speed by

v. The alien's speed is then $4v$. Denote Mulder's and alien's positions by M and A, respectively.

Mulder cannot just run for the fence along the straight line connecting C with the point on the fence diametrically opposite to A. Indeed, while it takes Mulder $\frac{R}{v}$ time to cover the distance R, the alien would cover the distance πR in $\frac{\pi R}{4v}$ time and the alien would catch up with Mulder, since $\frac{R}{v} > \frac{\pi R}{4v}$.

To optimize this strategy, Mulder needs to start running for the fence from a point that is closer to the fence than C. Assume that Mulder somehow managed to be at a point M that is xR away from C (where $0 < x < 1$), with M, C, and A collinear, and C between M and A. Denote by P the point on the circle diametrically opposite to A; see Figure 3.2. Then, $MC = xR$ and $MP = (1-x)R$. It takes Mulder $\frac{(1-x)R}{v}$ time to reach the fence running from M to P, while the alien needs $\frac{\pi R}{4v}$ time to reach the point P going from A to P on a semicircle. Note that

$$\frac{(1-x)R}{v} < \frac{\pi R}{4v}, \quad \text{if } x > 1 - \frac{\pi}{4}.$$

Thus, if $x > 1 - \frac{\pi}{4}$, Mulder would be able to escape the alien.

We are now ready to describe Mulder's escape strategy. He sets $x = 1 - \frac{\pi}{4} + 0.01$. Note that $x < \frac{1}{4}$, since $0.01 < \frac{\pi - 3}{4}$. Regardless of the alien's movement, Mulder first runs from C to any point on the circle of radius xR, centered at C. Then he runs around that circle, until his position M is such that M, C, and A are collinear, with C between M and A. He is able to do so, since $x < \frac{1}{4}$ and the alien's speed is only 4 times his speed.[23] Finally,

---

[23]Mulder is able to do so since, if $x < \frac{1}{4}$, his angular speed, $\frac{\theta v}{\pi R x}$, is larger than the angular speed $\frac{4\theta v}{\pi R}$ of the alien:

$$\frac{\theta v}{\pi R x} > \frac{4\theta v}{\pi R} \iff \frac{1}{4} > x.$$

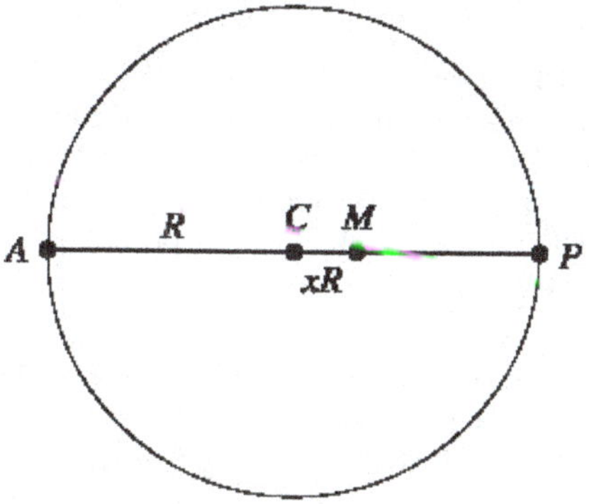

Figure 3.2: Mulder can reach P from M before the alien can do so from A.

Mulder runs from M to P and will reach P before alien does, as shown above, since $x > 1 - \frac{\pi}{4}$. □

Question 12. At your subway station, you notice that of the two trains running in opposite directions which are supposed to arrive with the same frequency, the train going in one direction comes first 80% of the time, while the train going in the opposite direction comes first only 20% of the time. What do you think could be happening?

Answer: One thing that could be happening is that the train that comes first 80% of the time comes in fact more frequently than the other one. However, even if both trains run with the same frequency, one train might come first 80% of the time. For example, assuming that your arrival in the station is uniformly distributed, if both trains run every ten minutes, and train A comes into the station at 1:00, 1:10, 1:20, ..., while train B comes at 1:12, 1:22, 1:32, ..., then train A will come first 80% of the time.

□

Question 13. You start off with one amoeba. Every minute, this amoeba can either die, do nothing, split into two amoebas, or split into three amoebas; all these scenarios being equally likely to happen. All further amoebas behave the same way. What is the probability that the amoebas eventually die off?

Answer:
Solution 1: (Due to Yu Gan, Baruch MFE'14.) Denote by A_k the event that no amoebas are alive after k minutes. Let $p_k = P(A_k)$. Note that $p_1 = \frac{1}{4}$ and $A_k \subseteq A_{k+1}$, for all $k \geq 1$.

The probability p that the amoebas eventually die off is the probability that at some point in time, i.e., after n minutes for some n, no amoebas are alive. In other words,

$$p = P\left(\bigcup_{k=1}^{\infty} A_k\right).$$

Note that $\bigcup_{k=1}^{n} A_k = A_n$, since $A_k \subseteq A_{k+1}$ for all $k \geq 1$. Then,

$$\begin{aligned} p &= P\left(\bigcup_{k=1}^{\infty} A_k\right) = P\left(\lim_{n\to\infty}\bigcup_{k=1}^{n} A_k\right) \\ &= \lim_{n\to\infty} P\left(\bigcup_{k=1}^{n} A_k\right) = \lim_{n\to\infty} P(A_n) \\ &= \lim_{n\to\infty} p_n. \end{aligned} \tag{3.299}$$

Given the four equally probable outcomes after the first minute, i.e, the amoeba can die, remain one amoeba, split into two amoebas, or split into three amoebas, it follows that

$$p_n = \frac{1}{4} + \frac{1}{4}p_{n-1} + \frac{1}{4}p_{n-1}^2 + \frac{1}{4}p_{n-1}^3, \tag{3.300}$$

for all $n \geq 2$.

The sequence $(p_n)_{n\geq 0}$ is increasing. Recall that $A_n \subseteq A_{n+1}$, and therefore

$$p_n = P(A_n) \leq P(A_{n+1}) = p_{n+1}, \quad \forall\, n \geq 1.$$

Also, we can see by induction that the sequence $(p_n)_{n\geq 0}$ is bounded from above by $\sqrt{2}-1$, since $p_1 = \frac{1}{4} < \sqrt{2}-1$, and, if we assume that $p_{n-1} < \sqrt{2}-1$ for some $n \geq 2$, then we obtain from (3.300) that

$$\begin{aligned} p_n &< \frac{1}{4} + \frac{1}{4}(\sqrt{2}-1) + \frac{1}{4}(\sqrt{2}-1)^2 + \frac{1}{4}(\sqrt{2}-1)^3 \\ &= \sqrt{2}-1. \end{aligned}$$

Thus, the sequence $(p_n)_{n\geq 0}$ is bounded from above and increasing, and therefore convergent.

Recall from (3.299) that $\lim_{n\to\infty} p_n = p$, where p is the probability that the amoebas eventually die off. Since $p_n < \sqrt{2}-1$ for all $n \geq 1$, it follows that $p \leq \sqrt{2}-1$. Moreover, from (3.300), we find that

$$p = \frac{1}{4} + \frac{1}{4}p + \frac{1}{4}p^2 + \frac{1}{4}p^3,$$

which can be written as

$$0 = p^3 + p^2 - 3p + 1 = (p-1)(p^2 + 2p - 1),$$

and has solutions 1, $\sqrt{2}-1$, and $-\sqrt{2}-1$. Since $0 < p \leq \sqrt{2}-1$, we obtain that $p = \sqrt{2}-1$.

In other words, the probability that the amoebas eventually die off is $\sqrt{2}-1$.

Solution 2: Let p be the probability that the descendants of a single amoeba will die out eventually. Then, the probability that the descendants of n amoebas will all die out eventually is p^n, since each amoeba is independent of all other amoebas.

Furthermore, the probability that the descendants of an amoeba will die out eventually is independent of time when averaged over all the possibilities.

At the beginning, the probability that the descendants of an amoeba will die out eventually is, by definition, p. After one minute, the initial amoeba turns into 0, 1, 2, or 3 amoebas, with probability of $\frac{1}{4}$ for each case. Thus, the probability that the descendants of the amoeba die out is now

$$\frac{1}{4}\left(p^0+p^1+p^2+p^3\right) \;=\; \frac{1}{4}\left(1+p+p^2+p^3\right).$$

These two probabilities must be equal, and therefore

$$p \;=\; \frac{1}{4}\left(1+p+p^2+p^3\right), \qquad (3.301)$$

which is the same as

$$p^3+p^2-3p+1 \;=\; (p-1)(p^2+2p-1) \;=\; 0. \quad (3.302)$$

The roots of (3.302) are $p = 1$, $p = -\sqrt{2}-1$, and $p = \sqrt{2}-1$. The only root in the interval $(0, 1)$ is $p = \sqrt{2}-1$, and this is the probability that the amoebas eventually die off.

A more subtle question is why $p = 1$ is not one of the possible answers to the problem in hand? The right hand side of (3.301) is the generating function $h(p)$ of amoeba's branching process. It is a well-known theorem on branching processes that if the mean number of offspring produced by a single amoeba is bigger than 1, then the smallest positive root of the equation $p = h(p)$ is the probability that amoeba's descendants will die out eventually. In our case, the mean number of offspring produced by a single amoeba is $\frac{1}{4}(0+1+2+3) = \frac{3}{2} > 1$, so the theorem applies. □

Question 14. Given a set X with n elements, choose two subsets A and B at random. What is the probability of A being a subset of B?

Answer: When two subsets A and B of X are chosen at random, each element of X is equally likely to end up in any of the following four sets:

$$A \setminus B, \quad B \setminus A, \quad A \cap B, \quad \text{and} \quad X \setminus (A \cup B).$$

For A to be a subset of B, $A \setminus B$ would have to be empty; in other words, none of the n elements of X would end up in $A \setminus B$. The probability of any element of X not ending up in $A \setminus B$ is $\frac{3}{4}$.

We conclude that the probability of A being a subset of B is

$$\left(\frac{3}{4}\right)^n. \quad \square$$

Question 15. Alice writes two distinct real numbers between 0 and 1 on two sheets of paper. Bob selects one of the sheets randomly to inspect it. He then has to declare whether the number he sees is the bigger or smaller of the two.

Is there any way Bob can expect to be correct more than half the times Alice plays this game with him?

Answer: Denote the numbers Alice writes on two sheets of paper by a_1 and a_2, $0 < a_1 < a_2 < 1$. Denote the number Bob selects by A. Bob's task is to guess (with probability of being correct bigger than 1/2) whether $A = a_1$ or $A = a_2$.

Bob's strategy is as follows: after seeing A, Bob draws a number B uniformly at random from $(0, 1)$; if B is smaller than A, Bob declares that $A = a_2$; otherwise, he declares that $A = a_1$.

Denote by E the event that Bob is correct using this strategy.

Then,

$$\begin{aligned} P(E) &= P(E|A = a_1) \cdot P(A = a_1) \\ &\quad + P(E|A = a_2) \cdot P(A = a_2) \\ &= P(B > a_1) \cdot P(A = a_1) \\ &\quad + P(B < a_2) \cdot P(A = a_2) \\ &= (1 - a_1) \cdot 0.5 + a_2 \cdot 0.5 \\ &= 0.5 + 0.5(a_2 - a_1) \\ &> 0.5. \end{aligned}$$

The probability that Bob is correct using this strategy is therefore greater than $\frac{1}{2}$. □

Question 16. How many digits does the number 125^{100} have? You are not allowed to use values of $\log_{10} 2$ or $\log_{10} 5$.

Answer: Note that

$$125^{100} = \left(\frac{1000}{8}\right)^{100} = \frac{1000^{100}}{2^{300}}. \quad (3.303)$$

Since $2^{10} = 1024$, we obtain that

$$125^{100} = \frac{1000^{100}}{1024^{30}} = \frac{1000^{70}}{1.024^{30}}. \quad (3.304)$$

We first show that

$$1 < 1.024^{30} < 10. \quad (3.305)$$

From the binomial expansion, it follows that

$$(1 + 0.024)^{30} = \sum_{j=0}^{30} \binom{30}{j} 0.024^j. \quad (3.306)$$

Note that the ratio of every two consecutive terms in (3.306) is less than 0.72, since

$$\begin{aligned}
\frac{\binom{30}{j+1}0.024^{j+1}}{\binom{30}{j}0.024^{j}} &= \frac{\binom{30}{j+1}}{\binom{30}{j}} \cdot 0.024 \\
&= \frac{30!}{(j+1)!\,(30-j-1)!} \cdot \frac{j!\,(30-j)!}{30!} \cdot 0.024 \\
&= \frac{30-j}{j+1} \cdot 0.024 \\
&< 30 \cdot 0.024 \\
&= 0.72.
\end{aligned}$$

Then,

$$\binom{30}{j} 0.024^j \;\leq\; 0.72^j \quad \forall\, 0 \leq j \leq 30, \tag{3.307}$$

and, from (3.306) and (3.307), we obtain that

$$\begin{aligned}
(1+0.024)^{30} &< \sum_{j=0}^{30} 0.72^j \;<\; \sum_{j=0}^{\infty} 0.72^j \\
&= \frac{1}{1-0.72} \\
&< 10;
\end{aligned}$$

note that, for the equality on the second line above, we used the geometric series identity

$$\sum_{j=1}^{\infty} z^j \;=\; \frac{1}{1-z} \quad \text{for} \quad z = 0.72.$$

Thus, the inequality (3.305) is proved.

From (3.304) and (3.305), we obtain that

$$10^{209} < 125^{100} = \frac{10^{210}}{1.024^{30}} < 10^{210}, \tag{3.308}$$

and conclude that 125^{100} has 210 digits. □

Question 17. For every subset of $\{1, 2, 3, \ldots, 2013\}$, arrange the numbers in the increasing order and take the sum with alternating signs. The resulting integer is called the weight of the subset.[24] Find the sum of the weights of all the subsets of $\{1, 2, 3, \ldots, 2013\}$.

Answer: Let $w(S)$ denote the weight of subset S. Every subset S of $\{1, 2, 3, \ldots, 2013\}$ that does not contain element 1 can be uniquely paired with the subset $\{1\} \cup S$ that contains element 1. Since there are 2^{2013} subsets of $\{1, 2, 3, \ldots, 2013\}$, there are 2^{2012} such pairs. Note that $w(S) + w(\{1\} \cup S) = 1$; that is, the combined weight of each pair is 1. For example,

$$\begin{aligned} & w(\{2,5,8\}) + w(\{1,2,5,8\}) \\ = & (2 - 5 + 8) + (1 - 2 + 5 - 8) \\ = & 1. \end{aligned}$$

Hence, the sum of the weights of all the subsets of $\{1, 2, 3, \ldots, 2013\}$ is 2^{2012}. □

Question 18. Alice and Bob alternately choose one number from one of the following nine numbers: 1/16, 1/8, 1/4, 1/2, 1, 2, 4, 8, 16, without replacement. Whoever gets three numbers that multiply to one wins the game. Alice starts first. What should her strategy be? Can she always win?

Answer: First, notice that the numbers Alice and Bob play with are powers of two, namely, 2^{-4}, 2^{-3}, 2^{-2}, 2^{-1}, 2^0, 2^1, 2^2, 2^3, and 2^4. Next, imagine that Alice and Bob are playing on the 3×3 square, whose entries are as follows: the first row (from left to right) is 2^3, 2^{-4}, 2^1; the

[24] For example, the weight of the subset $\{3\}$ is 3. The weight of the subset $\{2, 5, 8\}$ is $2 - 5 + 8 = 5$.

second row is 2^{-2}, 2^0, 2^2; and the third row is 2^{-1}, 2^4, 2^{-3}.

The product of the entries in every row, every column, and every diagonal is 1, and these possibilities cover all the ways to choose three of the given numbers multiplying to 1. Thus, Alice and Bob are essentially playing Tic-Tac-Toe! It is well-known that best play by both players in Tic-Tac-Toe leads to a draw. We conclude that Alice does not have a winning strategy, although she cannot lose either. □

Question 19. Mr. and Mrs. Jones invite four other couples over for a party. At the end of the party, Mr. Jones asks everyone else how many people they shook hands with, and finds that everyone gives a different answer. Of course, no one shook hands with his or her spouse and no one shook the same person's hand twice. How many people did Mrs. Jones shake hands with?

Answer: Since each person shook hands with at most eight others, the nine different answers received by Mr. Jones are exactly the numbers 0 through 8. Denote by P_i the person with i handshakes, $i = 0, 1, \ldots, 8$. Mr. Jones is not assigned any additional notation.

P_8 shook hands with 8 people of the total of 9 other people. Thus, P_8 did not shake the hand of only one other person, so that person must be his or her spouse. On the other hand, P_8 did not shake the hand of P_0 since nobody did that. Therefore, P_8 and P_0 are married, and P_8 shook everyone's hand except for P_0. P_7 did not shake the hands of two people, one of whom was his/her spouse. One of these two people had to be P_0 as he or she did not shake anyone's hand, and the other one had to be P_1 as he or she had only one handshake, namely with P_8. Since spouses do not shake hands, the spouse of P_7 is either P_1 or P_0. However, P_0 is married to P_8, so P_1 must be married to P_7.

Proceeding similarly, we find that P_6 and P_2 must be married, and that P_5 and P_3 must be married. Then, P_4 must be Mrs. Jones, since this is the only person whose spouse was not identified, and Mr. Jones was not any one of $P_0, \ldots, P_8$.

We conclude that Mrs. Jones shook hands with four people. □

Question 20. The New York Yankees and the San Francisco Giants are playing in the World Series (best of seven format). You would like to bet \$100 on the Yankees winning the World Series, but you can only place bets on individual games, and every time at even odds. How much should you bet on the first game?

Answer: Let P be a 5×5 matrix containing the net payoffs for all the states in this dynamic programming problem. More precisely, $P(i, j)$ denotes the net payoff in the state (i, j) when Yankees have won i and lost j games ($0 \leq i, j \leq 4$). Clearly, $P(4, j) = 100$ ($0 \leq j \leq 3$), and $P(i, 4) = -100$ ($0 \leq i \leq 3$). Moreover, $P(4, 4)$ is left blank, as $(4, 4)$ is not in our state space (4 wins and 4 losses cannot be achieved in a best of seven series).

Let B be a 4×4 matrix containing the bets we need to place at each state (i, j), given that we would like to bet \$100 on the Yankees winning the World Series, and given that Yankees have won i and lost j games so far ($0 \leq i, j \leq 3$). Clearly, $B(3, 3) = 100$.

Given that Yankees have won i and lost j games so far, if we bet $B(i, j)$ on the Yankees for the next game, our payoff will be $P(i, j) + B(i, j)$ if the Yankees win, or $P(i, j) - B(i, j)$ if the Yankees lose. Therefore,

$$P(i+1, j) = P(i, j) + B(i, j); \qquad (3.309)$$

$$P(i, j+1) = P(i, j) - B(i, j). \qquad (3.310)$$

By adding and subtracting the equations (3.309) and

(3.310) we obtain that

$$P(i,j) = \frac{1}{2}(P(i+1,j) + P(i,j+1)) \quad (3.311)$$

$$B(i,j) = \frac{1}{2}(P(i+1,j) - P(i,j+1)) \quad (3.312)$$

Now, it is easy to compute all the entries of P, using (3.311) and working backwards from $P(4,j)$ and $P(i,4)$. For example,

$$P(3,3) = \frac{1}{2}(P(4,3) + P(3,4)) = \frac{1}{2}(100 - 100) = 0.$$

Once matrix P is computed, we use (3.312) to compute the matrix B. For example,

$$B(2,1) = \frac{1}{2}(P(3,1) - P(2,2)) = \frac{1}{2}(75 - 0) = 37.5.$$

In other words, given that Yankees have won 2 games and lost 1 game so far, and we would like to bet \$100 on the Yankees winning the World Series, we should bet \$37.5 on the Yankees for the next game. We include both matrices below:

$$P = \begin{pmatrix} 0 & -31.25 & -62.5 & -87.5 & -100 \\ 31.25 & 0 & -37.5 & -75 & -100 \\ 62.5 & 37.5 & 0 & -50 & -100 \\ 87.5 & 75 & 50 & 0 & -100 \\ 100 & 100 & 100 & 100 & 0 \end{pmatrix};$$

$$B = \begin{pmatrix} 31.25 & 31.25 & 25 & 12.5 \\ 31.25 & 37.5 & 37.5 & 25 \\ 25 & 37.5 & 50 & 50 \\ 12.5 & 25 & 50 & 100 \end{pmatrix}.$$

Therefore, we should bet $B(1,1) = 31.25$ dollars (on the Yankees) on the first game. Note that the probability of the Yankees winning or losing a single game does not affect your betting strategy. □

Question 21. We have two red, two green and two yellow balls. For each color, one ball is heavy and the other is light. All heavy balls weigh the same. All light balls weigh the same. How many weighings on a scale are necessary to identify the three heavy balls?

Answer: It is clear that one weighing does not suffice. We show that the heavy balls can be identified in only two weighings.

Label the red balls R_1 and R_2, the green balls G_1 and G_2, and the yellow balls Y_1 and Y_2. Our first weighing is $\{R_1, G_1\}$ vs. $\{R_2, Y_1\}$.

If the scale is in balance, then either G_1 or Y_1 is heavy, but not both. Our second weighing is $\{G_1\}$ vs. $\{Y_1\}$. If G_1 is heavier, then the set of heavy balls is $\{R_2, G_1, Y_2\}$. If Y_1 is heavier, then the set of heavy balls is $\{R_1, G_2, Y_1\}$.

If $\{R_1, G_1\}$ is heavy, then either G_1 is heavy or Y_1 is light. Our second weighing is $\{G_1, Y_1\}$ vs. $\{G_2, Y_2\}$. If the scale is in balance, then G_1 is heavy; hence, the set of heavy balls is $\{R_1, G_1, Y_2\}$. If $\{G_1, Y_1\}$ is heavier, then G_1 and Y_1 are both heavy. The set of heavy balls is $\{R_1, G_1, Y_1\}$. If $\{G_2, Y_2\}$ is heavier, then G_2 and Y_2 are both heavy. The set of heavy balls is $\{R_1, G_2, Y_2\}$.

If $\{R_2, Y_1\}$ is heavy, then either Y_1 is heavy or G_1 is light. Our second weighing is $\{G_1, Y_1\}$ vs. $\{G_2, Y_2\}$. If the scale is in balance, then Y_1 is heavy; hence, the set of heavy balls is $\{R_2, G_2, Y_1\}$. If $\{G_1, Y_1\}$ is heavy, then G_1 and Y_1 are both heavy. The set of heavy balls is $\{R_2, G_1, Y_1\}$. If $\{G_2, Y_2\}$ is heavier, then G_2 and Y_2 are both heavy. The set of heavy balls is $\{R_2, G_2, Y_2\}$. □

Question 22. There is a row of 10 rooms and a treasure in one of them. Each night, a ghost moves the treasure to an adjacent room. You are trying to find the treasure, but can only check one room per day. How do you find it?

Answer: The treasure can be found in at most 16 days. Label the rooms 1 through 10. Denote by T_k the room where the treasure is on day k, and let R_k denote the room you check on day k. Adopt the following strategy: for days $k = 1, 2, \ldots, 8$, let $R_k = k + 1$; for days $k = 9, 10, \ldots, 16$, let $R_k = 18 - k$.

If T_1 is even, then we will find the treasure in one of the first eight days. In other words, there exists k with $1 \leq k \leq 8$ such that $T_k = R_k$. Note that for $1 \leq k \leq 8$, T_k and R_k have the same parity, since T_1 is even, $R_1 = 2$, and both T_k and R_k change by at most 1 from day to day, according to ghost's moves and our strategy. Hence, $T_k - R_k$ is even. Furthermore, $T_1 \neq 1$ and $T_8 \neq 10$, which implies $T_1 - R_1 \geq 0$ and $T_8 - R_8 \leq 0$. Since $T_k - R_k$ can change by at most 2 from day to day, there must exist some k, $1 \leq k \leq 8$, such that $T_k - R_k = 0$.

If T_1 is odd, then we claim $T_k = R_k$ for some k, $9 \leq k \leq 16$. Note that T_9 is odd since T_1 is odd and the treasure is moved to an adjacent room each night. Furthermore, for $9 \leq k \leq 16$, T_k and R_k have the same parity since T_9 is odd, $R_9 = 9$, and both T_k and R_k change by at most 1 from day to day, according to ghost's moves and our strategy. Hence, $R_k - T_k$ is even. Furthermore, $T_9 \neq 10$ and $T_{16} \neq 1$, which implies $R_9 - T_9 \geq 0$ and $R_{16} - T_{16} \leq 0$. Since $R_k - T_k$ can change by at most 2 from day to day, there must exist some k, $9 \leq k \leq 16$, such that $R_k - T_k = 0$.

Finally, note that this strategy can be generalized to any number $n \geq 2$ of rooms, when the treasure will be found in $2n - 4$ days, by checking the rooms $2, 3, \ldots, n-1, n-1, n-2, \ldots, 3, 2$, in that order. □

Question 23. How many comparisons do you need to find the maximum in a set of n distinct numbers? How many comparisons do you need to find both the maximum and minimum in a set of n distinct numbers?

Answer: We can find the maximum in $n-1$ comparisons as follows: let $\{x_1, x_2, \ldots, x_n\}$ denote the set of n distinct numbers. Scan the numbers from left to right, while maintaining the current maximum M. More precisely, set $x_1 = M$, and, in the ith comparison, compare M and x_{i+1}. If $M > x_{i+1}$, leave M as is; otherwise, set $M = x_{i+1}$. Do this for $i = 1, \ldots, n-1$.

Similarly, one can find the minimum in a set of n distinct numbers in $n-1$ comparisons.

Then, we can find the maximum M and the minimum m in $\{x_1, x_2, \ldots, x_n\}$ with $2n-3$ comparisons: find the maximum M in $\{x_1, x_2, \ldots, x_n\}$ with $n-1$ comparisons, and then find the minimum m in $\{x_1, x_2, \ldots, x_n\} \setminus \{M\}$ with $n-2$ comparisons.

However, one can find m and M significantly faster.

If n is even, compare all $\frac{n}{2}$ consecutive pairs of numbers x_{2i-1} and x_{2i}, $i = 1 : \frac{n}{2}$, and put the smaller number into a set S and the larger number into a set L. This requires $\frac{n}{2}$ comparisons. Note that S and L have $\frac{n}{2}$ elements each. Then, find the minimum m in S using $\frac{n}{2} - 1$ comparisons, and find the maximum M in L using $\frac{n}{2} - 1$ comparisons. The total number of comparisons to find m and M is

$$\frac{n}{2} + \left(\frac{n}{2} - 1\right) + \left(\frac{n}{2} - 1\right) = \frac{3n}{2} - 2. \qquad (3.313)$$

If n is odd, compare $\frac{n-1}{2}$ consecutive pairs of numbers x_{2i-1} and x_{2i}, $i = 1 : \frac{n-1}{2}$, and put the smaller number into a set S and the larger number into a set L. This requires $\frac{n-1}{2}$ comparisons. Place x_n into both S and L. Note that S and L have $\frac{n-1}{2} + 1 = \frac{n+1}{2}$ elements each. Then, find the minimum m in S using $\frac{n+1}{2} - 1 = \frac{n-1}{2}$ comparisons, and find the maximum M in L using $\frac{n+1}{2} - 1 = \frac{n-1}{2}$ comparisons. The total number of comparisons to find m and M is

$$\frac{n-1}{2} + \frac{n-1}{2} + \frac{n-1}{2} = \frac{3n+1}{2} - 2. \qquad (3.314)$$

Note that the results of (3.313) and (3.314) can be written succinctly as

$$\left\lceil \frac{3n}{2} \right\rceil - 2 \quad \text{comparisons,}$$

where $\lceil x \rceil$ denotes the ceiling of x, the smallest integer greater than or equal to x. □

Question 24. Given a cube, you can jump from one vertex to a neighboring vertex with equal probability. Assume you start from a certain vertex (does not matter which one). What is the expected number of jumps to reach the opposite vertex?

Answer: Label the vertices of the cube with 0, 1, 2, and 3, according to your distance from the opposite vertex. In other words, label the starting vertex with 3, the vertices adjacent to the starting vertex with 2, the vertices adjacent to the opposite vertex with 1, and the opposite vertex with 0. Call the opposite vertex your final destination.

Denote by E_i, $i = 0, 1, 2, 3$, the expected number of jumps yet to be made to reach the final destination, given that you are currently in one of the vertices labeled with i. Note that $E_0 = 0$, and we have to find E_3.

After the first jump, you are in one of the vertices labeled with 2, so

$$E_3 = 1 + E_2. \qquad (3.315)$$

From a vertex labeled with 2, you can jump to three vertices: two of them are labeled with 1, and one of them is labeled with 3. Thus, you jump to a vertex labeled with 1 with probability $\frac{2}{3}$, or to a vertex labeled with 3 with probability $\frac{1}{3}$. Hence,

$$E_2 = 1 + \frac{2}{3}E_1 + \frac{1}{3}E_3. \qquad (3.316)$$

Similarly, from one of the vertices labeled with 1, you jump to a vertex labeled with 2 with probability $\frac{2}{3}$, or to a vertex labeled with 0 with probability $\frac{1}{3}$. Hence,

$$E_1 = 1 + \frac{2}{3}E_2 + \frac{1}{3}E_0 = 1 + \frac{2}{3}E_2, \qquad (3.317)$$

since $E_0 = 0$.

Solving (3.315–3.317) yields $E_1 = 7$, $E_2 = 9$, and $E_3 = 10$.

We conclude that it will take 10 jumps, on average, to reach the opposite vertex. □

Question 25. Select numbers uniformly distributed between 0 and 1, one after the other, as long as they keep decreasing; i.e. stop selecting when you obtain a number that is greater than the previous one you selected.

(i) On average, how many numbers have you selected?

(ii) What is the average value of the smallest number you have selected?

Answer: We give three solutions for part (i); the third solution will be used to solve part (ii).

(i) *Solution 1:* Denote by $E(x)$ the expected number of numbers you have yet to select, given that you have just selected number x. For example, $E(0) = 1$, since the next number you select is greater than $x = 0$, upon which the game stops.

Assume that you have just selected number x. Denote by y the next number you select. We find $E(x)$ by conditioning on y. With probability $1 - x$, y is greater than x; the game stops, and, thus, you have selected only one number after selecting x. On the other hand, y could be smaller than x, in which case you expect to select $E(y)$ additional numbers after selecting y; in other words, you expect to select $1+E(y)$ additional numbers after selecting x. Since the probability density function of y is $f(y) = 1$, the law of total probability gives

$$\begin{aligned} E(x) &= 1\cdot(1-x) + \int_0^x (1+E(y))f(y)\,dy \\ &= (1-x) + \int_0^x (1+E(y))\,dy \\ &= 1 + \int_0^x E(y)\,dy. \end{aligned}$$

Differentiating

$$E(x) = 1 + \int_0^x E(y)\,dy$$

with respect to x yields

$$E'(x) = E(x).$$

Thus, $E(x) = Ce^x$, where C is a constant. The condition $E(0) = 1$ gives $C = 1$. Hence, $E(x) = e^x$.

Note that the first number selected is automatically smaller than 1. Then, the number of numbers selected after starting with the number 1, which was denoted by $E(1)$, is equal to the number of numbers selected starting with a random number between 0 and 1. Therefore, the average number of numbers you have selected is $E(1) = e$.

(i) *Solution 2:* Denote by N the average number of numbers you have selected. Denote by x_i the ith number you selected, $i \geq 1$, and let p_i denote the probability that $x_i < x_{i-1} < \ldots < x_1$. Since there are $i!$ permutations of $x_1, \ldots, x_i$, then $p_i = \frac{1}{i!}$.

You select at least two numbers before stopping. The probability that you select exactly i numbers, $i \geq 2$, before stopping is equal to the probability that $x_{i-1} < \ldots < x_2 < x_1$ and $x_i > x_{i-1}$. The latter equals the probability that $x_{i-1} < \ldots < x_2 < x_1$ minus the probability that $x_i < x_{i-1} < \ldots < x_2 < x_1$; that is, $p_{i-1} - p_i$.

We conclude that the expected number of numbers you have selected is

$$\begin{aligned} N &= \sum_{i=2}^{\infty} i(p_{i-1} - p_i) \\ &= \sum_{i=2}^{\infty} i\left(\frac{1}{(i-1)!} - \frac{1}{i!}\right) \\ &= \sum_{i=2}^{\infty} i \cdot \frac{i-1}{i!} \\ &= \sum_{i=2}^{\infty} \frac{1}{(i-2)!} \\ &= \sum_{j=0}^{\infty} \frac{1}{j!} \\ &= e, \end{aligned} \tag{3.318}$$

where (3.318) follows from the Taylor series expansion of e^x around 0, i.e.

$$e^x = \sum_{j=0}^{\infty} \frac{x^j}{j!}, \quad \forall\, x \in \mathbb{R},$$

by letting $x = 0$.

(i) *Solution 3:* Denote by $p(x)\,dx$ the probability that a number between x and $x + dx$ is selected as part of the decreasing sequence. Let $p_i(x)\,dx$ denote the probability that a number between x and $x + dx$ is selected as the ith term of the decreasing sequence. Then,

$$p(x)\,dx = \left(\sum_{i=1}^{\infty} p_i(x)\right) dx. \tag{3.319}$$

The probability that a number between x and $x + dx$ is selected as the ith term of the decreasing sequence is

$$p_i(x)\,dx = \frac{1}{(i-1)!}(1-x)^{i-1}\,dx, \tag{3.320}$$

since $(1-x)^{i-1}$ is the probability that the first $i-1$ numbers selected are greater than x, and $\frac{1}{(i-1)!}$ is the probability that they are selected in decreasing order.

From (3.319) and (3.320), we obtain that

$$\begin{aligned} p(x)\,dx &= \sum_{i=1}^{\infty} \frac{(1-x)^{i-1}}{(i-1)!}\,dx \\ &= \sum_{j=0}^{\infty} \frac{(1-x)^j}{j!}\,dx \\ &= e^{1-x}\,dx; \qquad (3.321) \end{aligned}$$

where (3.321) follows from the Taylor series expansion of e^t around 0, i.e.

$$e^t = \sum_{j=0}^{\infty} \frac{t^j}{j!}, \quad \forall\, t \in \mathbb{R},$$

by letting $t = 1 - x$.

Therefore, the expected number of numbers selected in the decreasing sequence is

$$\int_0^1 p(x)\,dx = \int_0^1 e^{1-x}\,dx = e - 1.$$

Adding the last number selected (which is not in the decreasing sequence) gives an average of e numbers selected.

(ii) Denote by s the smallest number selected. It is the last number selected in the decreasing sequence. Since the probability that a number between x and $x + dx$ is selected as part of the decreasing sequence equals $e^{1-x}\,dx$, see (3.321), and since the probability that the next number selected is larger is $(1-x)$, then the probability that s is between x and $x + dx$ is $e^{1-x}(1-x)\,dx$.

Therefore, the expected value of the smallest number you have selected is

$$
\begin{aligned}
E(s) &= \int_0^1 xe^{1-x}(1-x)\,dx \\
&= \int_0^1 e^y y(1-y)\,dy, \qquad (3.322)
\end{aligned}
$$

where (3.322) follows from the substitution $y = 1 - x$.

Using integration by parts to compute (3.322), we find that

$$
\begin{aligned}
E(s) &= e^y(y-y^2)\Big|_0^1 - \int_0^1 e^y(1-2y)\,dy \\
&= e^y(y-y^2)\Big|_0^1 - e^y\Big|_0^1 + 2\int_0^1 ye^y\,dy \\
&= e^y(y-y^2-1)\Big|_0^1 + 2ye^y\Big|_0^1 - 2\int_0^1 e^y\,dy \\
&= e^y(3y-y^2-1)\Big|_0^1 - 2e^y\Big|_0^1 \\
&= e^y(3y-y^2-3)\Big|_0^1 \\
&= 3-e. \quad \square
\end{aligned}
$$

Question 26. To organize a charity event that costs \$100K, an organization raises funds. Independent of each other, one donor after another donates some amount of money that is exponentially distributed with a mean of \$20K. The process is stopped as soon as \$100K or more has been collected. Find the distribution, mean, and variance of the number of donors needed until at least \$100K has been collected.

Answer: Denote by a_i the amount of money donated by donor i, that is exponentially distributed with mean $1/\lambda$. Let $s_n = \sum_{i=1}^n a_i$ be the total amount of money donated by donors $1, \ldots, n$, and let

$$N = \min_{n \geq 1} \{n \text{ such that } s_n \geq a\}$$

be the discrete random variable denoting the smallest index n such that s_n is at least a.

Denote by $P(n|a)$, $n \geq 1$, the probability mass function of N, that is, the probability that $N = n$ when a total of a needs to be raised. Note that

$$P(1|a) = P(a_1 \geq a) = e^{-\lambda a}. \tag{3.323}$$

We find $P(n|a)$, $n > 1$, by conditioning on a_1. Given that the first donor donated $a_1 = x < a$, N is equal to n if and only if the remaining amount $a - x$ is raised by the next $n - 1$ donors (and not by fewer than the next $n - 1$ donors), an event that by definition has probability $P(n-1|a-x)$. Since the probability density function of a_1 is $f_{a_1}(x) = \lambda e^{-\lambda x}$, then, for $n > 1$, the law of total probability yields

$$\begin{aligned} P(n|a) &= \int_0^a P(n-1|a-x) f_{a_1}(x)\, dx \\ &= \int_0^a \lambda e^{-\lambda x} P(n-1|a-x)\, dx. \end{aligned} \tag{3.324}$$

We will prove that

$$P(n|a) = \frac{(\lambda a)^{n-1}}{(n-1)!} e^{-\lambda a}, \quad \forall\, a \geq 0, \tag{3.325}$$

by induction on n. The base case $n = 1$ was already established; see (3.323). Assume that (3.325) holds for $n > 1$; we will show that it also holds for $n + 1$.

From the induction hypothesis, we obtain that

$$P(n|a-x) = \frac{(\lambda(a-x))^{n-1}}{(n-1)!} e^{-\lambda(a-x)}, \tag{3.326}$$

for all $0 \le x \le a$. From (3.324), it follows that

$$P(n+1|a) = \int_0^a \lambda e^{-\lambda x} P(n|a-x)\,dx. \qquad (3.327)$$

From (3.326) and (3.327), we find that

$$\begin{aligned} & P(n+1|a) \\ = \; & \int_0^a \lambda e^{-\lambda x} \cdot \frac{(\lambda(a-x))^{n-1}}{(n-1)!}\, e^{-\lambda(a-x)}\,dx \\ = \; & \frac{\lambda^n e^{-\lambda a}}{(n-1)!} \int_0^a (a-x)^{n-1}\,dx \\ = \; & \frac{(\lambda a)^n}{n!}\, e^{-\lambda a}. \end{aligned}$$

We conclude that (3.325) holds for $n+1$, and therefore (3.325) is proved by induction.

From (3.325), it follows that N has the same distribution as $1 + M$, where M has a Poisson distribution with mean λa. Then,

$$E[N] = 1 + \lambda a; \quad \mathrm{Var}(N) = \lambda a.$$

For our problem, $1/\lambda = \$20\mathrm{K}$ and $a = \$100\mathrm{K}$. Thus, $E[N] = 1 + \lambda a = 6$ and $\mathrm{Var}(N) = \lambda a = 5$.

We conclude that the number of donors needed until at least \$100K is collected has mean 6 and variance 5. □

Question 27. Consider a random walk starting at 1 and with equal probability of moving to the left or to the right by one unit, and stopping either at 0 or at 3.

(i) What is the expected number of steps to do so?

(ii) What is the probability of the random walk ending at 3 rather than at 0?

Answer: Denote by X_n^ℓ the position of the random walk at time n, where the superscript refers to the starting position of the walk; for example, $X_0^\ell = \ell$. In this question, we are concerned with X_n^1. For a random walk starting at $\ell \in \{0, 1, 2, 3\}$, denote by T_ℓ the number of steps taken by the random walk in order to reach either 0 or 3 for the first time, and by $t_\ell = E[T_\ell]$ the expected number of steps.

(i) We are looking to find t_1. We derive a recurrence relation among the t_ℓ's as follows: for $\ell = 1, 2$,

$$\begin{aligned} t_\ell &= E[T_\ell] \\ &= E[T_\ell | X_1^\ell = \ell - 1] \cdot P(X_1^\ell = \ell - 1) \\ &\quad + E[T_\ell | X_1^\ell = \ell + 1] \cdot P(X_1^\ell = \ell + 1) \\ &= E[1 + T_{\ell-1}] \cdot P(X_1^\ell = \ell - 1) \qquad (3.328) \\ &\quad + E[1 + T_{\ell+1}] \cdot P(X_1^\ell = \ell + 1) \qquad (3.329) \\ &= (1 + t_{\ell-1}) \cdot \frac{1}{2} + (1 + t_{\ell+1}) \cdot \frac{1}{2}, \qquad (3.330) \end{aligned}$$

where the following identities were used to derive (3.328) and (3.329):

$$E[T_\ell | X_1^\ell = \ell - 1] = E[1 + T_{\ell-1}]; \qquad (3.331)$$

$$E[T_\ell | X_1^\ell = \ell + 1] = E[1 + T_{\ell+1}]. \qquad (3.332)$$

Note that (3.331) and (3.332) follow from the fact that, starting at ℓ, once the random walk took its first step to $\ell - 1$ (or $\ell + 1$), it becomes equivalent to a random walk starting afresh at $\ell - 1$ (or $\ell + 1$). The plus 1 term on the right hand side of (3.331) and (3.332) accounts for the first step.

Since $t_0 = t_3 = 0$, by letting $l = 1$ and then $l = 2$ in (3.330), we obtain the following linear system for t_1 and t_2:

$$\begin{aligned} t_1 &= \frac{1 + t_0}{2} + \frac{1 + t_2}{2} \\ &= \frac{1}{2} + \frac{1 + t_2}{2}; \end{aligned}$$

$$\begin{aligned} t_2 &= \frac{1+t_1}{2} + \frac{1+t_3}{2} \\ &= \frac{1+t_1}{2} + \frac{1}{2}. \end{aligned}$$

Thus, $t_1 = 2$ and $t_2 = 2$.

We conclude that the expected number of steps before stopping is 2.

(ii) Denote by p_ℓ the probability that the random walk reaches 3 before it reaches 0 when its position is at ℓ, for $\ell \in \{0,1,2,3\}$. We need to find p_1. Denote by τ_0^ℓ the first time the walk reaches 0 when starting at ℓ, and by τ_3^ℓ the first time the walk reaches 3 when starting at ℓ. We derive a recurrence relation for the p_ℓ's as follows: for $\ell = 1, 2$,

$$\begin{aligned} p_\ell &= P\left(\tau_3^\ell < \tau_0^\ell\right) \\ &= P\left(\tau_3^\ell < \tau_0^\ell \middle| X_1^\ell = \ell+1\right) P(X_1^\ell = \ell+1) \\ &\quad + P\left(\tau_3^\ell < \tau_0^\ell \middle| X_1^\ell = \ell-1\right) P(X_1^\ell = \ell-1) \\ &= P\left(\tau_3^{\ell+1} < \tau_0^{\ell+1}\right) P(X_1^\ell = \ell+1) \\ &\quad + P\left(\tau_3^{\ell-1} < \tau_0^{\ell-1}\right) P(X_1^\ell = \ell-1) \\ &= \frac{1}{2} \cdot p_{\ell+1} + \frac{1}{2} \cdot p_{\ell-1}, \end{aligned} \tag{3.333}$$

because the random walk is equally likely to move to the left or to the right and, once it moved, the random walk starts afresh.

Since $p_0 = 0$ and $p_3 = 1$, by letting $l = 1$ and then $l = 2$ in (3.333), we obtain the following linear system for p_1 and p_2:

$$\begin{aligned} p_1 &= \frac{p_0}{2} + \frac{p_2}{2} = \frac{p_2}{2}; \\ p_2 &= \frac{p_1}{2} + \frac{p_3}{2} = \frac{p_1}{2} + \frac{1}{2}. \end{aligned}$$

Thus, $p_1 = \frac{1}{3}$ and $p_2 = \frac{2}{3}$.

We conclude that the probability of the random walk ending at 3 rather than at 0 is $\frac{1}{3}$. □

Question 28. A stick of length 1 drops and breaks at a random place uniformly distributed across the length. What is the expected length of the smaller part?

Answer:

Solution 1: Treating the stick as an interval $[0, 1]$, the breakpoint X becomes a random variable uniformly distributed on $(0, 1)$. Its probability density function $f_X(x)$ is 1, for $0 \leq x \leq 1$, and 0 otherwise. Denote by L the length of the smaller part. Then, $L = \min\,(x, 1-x)$. We conclude that

$$\begin{aligned} E[L] &= \int_0^1 \min\,(x, 1-x) \cdot f_X(x)\, dx \\ &= \int_0^1 \min\,(x, 1-x)\, dx \\ &= \int_0^{1/2} x\, dx + \int_{1/2}^1 (1-x)\, dx \\ &= \frac{1}{8} + \frac{1}{8} \\ &= \frac{1}{4}. \end{aligned}$$

Solution 2: Treating the stick as an interval $[0, 1]$, the breakpoint X becomes a random variable uniformly distributed on $(0, 1)$. Denote by L the length of the smaller part.

Let A be the event that the breakpoint X is in $\left(0, \frac{1}{2}\right)$. Then, $\overline{A}$ is the event that the breakpoint X is in $\left(\frac{1}{2}, 1\right)$. Clearly, $P(A) = P(\overline{A}) = \frac{1}{2}$. Given A, the length of the smaller part is X. Given $\overline{A}$, the length of the smaller part is $1 - X$.

The law of total probability yields

$$\begin{aligned} E[L] &= E[L|A] \cdot P(A) + E[L|\overline{A}] \cdot P(\overline{A}) \\ &= E[X|A] \cdot \frac{1}{2} + E[1 - X|\overline{A}] \cdot \frac{1}{2} \\ &= E[X|A] \cdot \frac{1}{2} + \frac{1}{2} - E[X|\overline{A}] \cdot \frac{1}{2}. \quad (3.334) \end{aligned}$$

Note that, given A, X is uniformly distributed on $\left(0, \frac{1}{2}\right)$, and, Given $\overline{A}$, X is uniformly distributed on $\left(\frac{1}{2}, 1\right)$. Since the expectation of a random variable uniformly distributed on an interval (a, b) is equal to $\frac{a+b}{2}$, we obtain that

$$E[X|A] = \frac{1}{4} \quad \text{and} \quad E[X|\overline{A}] = \frac{3}{4}.$$

Then, (3.334) yields

$$\begin{aligned} E[L] &= \frac{1}{4} \cdot \frac{1}{2} + \frac{1}{2} - \frac{3}{4} \cdot \frac{1}{2} \\ &= \frac{1}{4}. \quad \square \end{aligned}$$

Question 29. You are given a stick of unit length.

(i) The stick drops and breaks at two places. What is the probability that the three pieces could form a triangle?

(ii) The stick drops and breaks at one place. Then the larger piece is taken and dropped again, breaking at one place. What is the probability that the three pieces could form a triangle?

Answer: We offer two solutions for part (i); the second solution will be used to solve part (ii).

(i) *Solution 1:* Denote by X and Y the two break points, and assume X and Y are independent random variables uniformly distributed on $(0, 1)$. To form a triangle, the sum of the lengths of any two pieces must be greater than the length of the third piece. Equivalently, each piece must be of length less than 1/2.

Assume that $Y > X$. Then, the length of the three pieces are X, $Y - X$, and $1 - Y$. Each of these pieces is of length less than 1/2 if and only if the point (X, Y) belongs to the region $\{(x, y) : x < 1/2, y - x < 1/2, 1 - y < 1/2, x \in (0, 1), y \in (0, 1), x < y\}$ in the unit square. Since the area of this region is 1/8, the probability of the three pieces forming a triangle, given $Y > X$, is 1/8. Symmetrically, the probability of the three pieces forming a triangle, given $Y < X$, is 1/8. Events $\{Y > X\}$ and $\{Y < X\}$ are disjoint; hence, the probability of the three pieces forming a triangle is 1/4.

(i) *Solution 2:* Consider an equilateral triangle ABC with height of length 1. Given a point P in its interior, let h_a, h_b, and h_c be the lengths of the perpendiculars dropped from P to the sides BC, CA, and AB, respectively. Since the areas of triangles BPC, CPA, and APB sum up to the area of ABC, we conclude that $h_a + h_b + h_c = 1$, and, thus, is independent of the position of P. Breaking a stick of length 1 into three pieces of lengths h_a, h_b, and h_c, is clearly equivalent to (uniquely) specifying a point P in the interior of the triangle ABC.

Connect the midpoints A', B', C', of the sides of triangle ABC to split it into four congruent equilateral triangles, with the medial triangle $A'B'C'$ in the middle (see Figure 3.3). Each piece of the broken stick has length less than 1/2 if and only if the corresponding point P belongs to the medial triangle. Since the area of the medial triangle is 1/4 of the area of triangle ABC, the desired probability is 1/4.

(ii) Assume that the pieces have lengths h and $(1 - h)$ after the first break, with $h < (1 - h)$, i.e., $h < 1/2$. With h fixed, the (larger) piece of length $1 - h$ is taken and dropped again, breaking at one place uniformly at random. The probability that the three pieces thus obtained

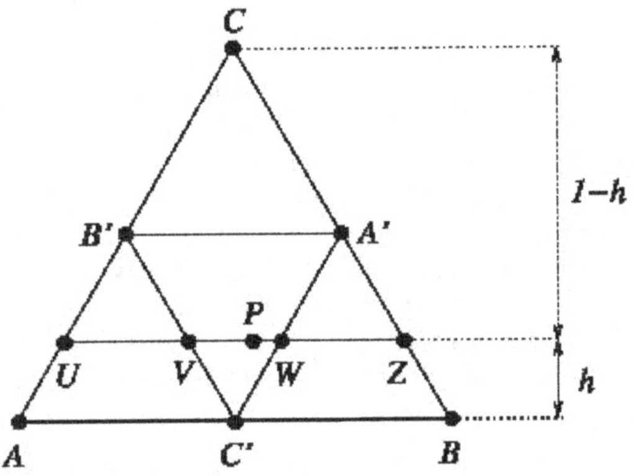

Figure 3.3: Point P inside the medial triangle $A'B'C'$ of the equilateral triangle ABC.

form a triangle clearly depends on h. It is close to 0 for h close to 0, and it is close to 1 for h close to 1/2.

More precisely, using the representation from the second solution of part (a) above, the probability that the three pieces form a triangle, given a fixed $h < 1/2$, is equal to the probability that the point P, that lies on the segment UZ parallel to side AB and at distance h from it, belongs to the segment VW, that is the intersection of UZ with the medial triangle $A'B'C'$ (see Figure 3.3). Since P is chosen (by the second break of the stick) from UZ uniformly at random, the probability that P belongs to VW is equal to the ratio of their lengths, namely $\frac{VW}{UZ}$.

Next, we express this ratio in terms of h. First, note that the medial triangle $A'B'C'$ has side length $AB/2$ and height 1/2. Since the triangles ABC and UZC are similar, we have $\frac{UZ}{AB} = \frac{1-h}{1}$. Since the triangles WVC'

and $A'B'C'$ are similar, we have $\frac{VW}{A'B'} = \frac{h}{1/2}$. Dividing the last two equations, we obtain that

$$\frac{VW}{UZ} = \frac{h}{1-h}.$$

Therefore, given $h < 1/2$, the probability that the three pieces form a triangle is

$$\int_0^{1/2} \frac{h}{1-h}\,dh = -\int_0^{1/2} \frac{1-h}{1-h}\,dh + \int_0^{1/2} \frac{1}{1-h}\,dh.$$

By using the substitution $y = 1 - h$, it follows that

$$\begin{aligned}\int_0^{1/2} \frac{h}{1-h}\,dh &= -\frac{1}{2} + \int_{1/2}^{1} \frac{1}{y}\,dy \\ &= -\frac{1}{2} - \ln\left(\frac{1}{2}\right) \\ &= \ln 2 - \frac{1}{2}.\end{aligned}$$

Since the probability that $h < 1/2$, where h is chosen uniformly at random from $(0,1)$ (by the first break of the stick), is $1/2$, then the total probability that the three pieces form a triangle is

$$\frac{\ln 2 - 1/2}{1/2} = 2\ln 2 - 1. \quad \square$$

Question 30. Why is a manhole cover round?

Answer: A circle is the shape with minimal surface given a required minimal width in any direction. Moreover, the cover of a round manhole cannot fall through the hole. If the manhole were square, its cover turned on its edge could fall through the hole since the diagonal of a square is $\sqrt{2}$ times larger than its edge. $\square$

Question 31. When is the first time after 12 o'clock that the hour and minute hands of a clock meet again?

Answer: The minute hand moves at a speed of 360 degrees per hour, while the hour hand moves at a speed of 30 degrees per hour. They start together at 12 o'clock. The first time they meet, the minute hand made one full rotation more than the hour hand, which is the same as 360 degrees more than the hour hand. If t denotes the time (measured in hours) until the two hands meet again, this can be written as

$$360 \cdot t = 30 \cdot t + 360.$$

Thus, $t = \frac{12}{11}$ hours, which is approximately 1 hour, 5 minutes, and 27 seconds. □

Question 32. Three light switches are in one room, and they turn three light bulbs in another. How do you figure out which switch turns on which bulb in one shot?

Answer: Turn on two switches for a couple of minutes, and then turn one of the switches off and go into the other room. The bulb that is lit corresponds to the switch that is still on; the bulb that is not lit but is hot corresponds to the switch that was turned on and then turned off; the bulb that is not lit and is cold corresponds to the switch that was never turned on. □

Question 33. The number 2^{29} has 9 digits, all different. Without computing 2^{29}, find the missing digit.

Answer: For any positive integer n, denote by $D(n)$ the sum of the digits of n. Recall that the difference between a number and the sum of its digits is divisible by 9, i.e.,

$$9 \mid n - D(n).$$

Thus, for $n = 2^{29}$, it follows that

$$9 \mid 2^{29} - D\left(2^{29}\right). \tag{3.335}$$

We are given that 2^{29} has 9 digits, and that all 9 digits are different. Denote by x the missing digit. Then,

$$D\left(2^{29}\right) = \left(\sum_{j=0}^{9} j\right) - x = 45 - x. \qquad (3.336)$$

From (3.335) and (3.336), it follows that

$$9 \mid 2^{29} - (45 - x). \qquad (3.337)$$

Note that

$$\begin{aligned} 2^{29} &= 2^5 \cdot (2^6)^4 = 2^5 \cdot 64^4 \\ &= 2^5 \cdot (63 + 1)^4 \\ &= 2^5 \cdot (63 \cdot k + 1) \\ &= 2^5 \cdot 63 \cdot k + 2^5, \end{aligned} \qquad (3.338)$$

where k is a positive integer.[25]

From (3.338), we find that

$$2^{29} - 2^5 = 63 \cdot 2^5 \cdot k,$$

and therefore

$$9 \mid 2^{29} - 2^5. \qquad (3.339)$$

From (3.337) and (3.339), it follows that

$$\begin{aligned} 9 \mid\ & (2^{29} - 2^5) - (2^{29} - (45 - x)) \\ & = (45 - x) - 2^5 \\ & = 13 - x. \end{aligned}$$

[25]It is easy to see that

$$\begin{aligned} (63 + 1)^4 &= 63^4 + 4 \cdot 63^3 + 6 \cdot 63^2 + 4 \cdot 63 + 1 \\ &= 63 \cdot (63^3 + 4 \cdot 63^2 + 6 \cdot 63 + 4) + 1 \\ &= 63 \cdot k + 1, \end{aligned}$$

where $k = 63^3 + 4 \cdot 63^2 + 6 \cdot 63 + 4$.

Since $9 \mid 13 - x$ and x is a digit, we conclude that $x = 4$. In other words, we identified that x, the missing digit from 2^{29}, must be 4.

Indeed, $2^{29} = 536\,870\,912$, i.e., 2^{29} has 9 digits, all different, and 4 is not a digit of 2^{29}.

For completeness, we include here a proof of the fact that the difference between a number and the sum of its digits is divisible by 9, i.e.,

$$9 \mid n - D(n).$$

If the digits of n are $a_k, a_{k-1}, \ldots a_1, a_0$ (from left to right), then

$$\begin{aligned} n &= a_k \cdot 10^k + a_{k-1} \cdot 10^{k-1} + \ldots a_1 \cdot 10 + a_0; \\ D(n) &= a_k + a_{k-1} + \ldots + a_1 + a_0. \end{aligned}$$

Hence,

$$n - D(n) = \sum_{i=0}^{k} a_i \cdot (10^i - 1).$$

Since $10^i - 1$ is an i-digit number with all digits equal to 9, it follows that $9 \mid 10^i - 1$, for all $i = 1 : k$, and therefore $9 \mid n - D(n)$. $\square$

Question 34. Alice and Bob stand at opposite ends of a straight line segment. Bob sends 50 ants towards Alice, one after another. Alice sends 20 ants towards Bob. All ants travel along the straight line segment. Whenever two ants collide, they simply bounce back and start traveling in the opposite direction. How many ants reach Bob and how many ants reach Alice? How many ant collisions take place?

Answer: Imagine that when two ants meet, they switch identities. Hence, even after a collision, two ants are traveling in two opposite directions. It follows that 20 ants reach Bob, while 50 ants reach Alice.

To calculate the number of ant collisions, imagine that each ant carries a message. In other words, Bob sends 50 messages to Alice, one message per ant. Similarly, Alice sends 20 messages to Bob, one message per ant. Furthermore, imagine that the two ants swap messages when they collide. Then a message always makes forward progress. Each of Alice's messages goes through 50 ant collisions. Each of Bob's messages goes through 20 ant collisions. The total number of collisions is 50 times 20, which is 1000 collisions. □

Question 35. There are 20 people at a party. Everyone writes down their name on a piece of paper and throws it in a bag. We shake up the bag and each person draws one name from the bag. You are in the same group as the person you have drawn.
For example, if people labeled 1 through 20 drew the following names from the bag:

$$\begin{array}{cccccccccc} 1 & 2 & 3 & 4 & 5 & 6 & 7 & 8 & 9 & 10 \\ \downarrow & \downarrow & \downarrow & \downarrow & \downarrow & \downarrow & \downarrow & \downarrow & \downarrow & \downarrow \\ 6 & 5 & 3 & 8 & 11 & 10 & 9 & 13 & 15 & 12 \end{array}$$

$$\begin{array}{cccccccccc} 11 & 12 & 13 & 14 & 15 & 16 & 17 & 18 & 19 & 20 \\ \downarrow & \downarrow & \downarrow & \downarrow & \downarrow & \downarrow & \downarrow & \downarrow & \downarrow & \downarrow \\ 14 & 18 & 4 & 20 & 17 & 16 & 7 & 19 & 1 & 2 \end{array}$$

the groups that form are

$$(1, 6, 10, 12, 18, 19), (2, 5, 11, 14, 20), (3),$$

$$(4, 8, 13), (7, 9, 15, 17), (16).$$

What is the expected number of groups?
Answer:
Solution 1: (Due to Zhaofeng Brent Liao, Baruch MFE'19.) Consider the general case when there are n people at the party. Let the random variable X_n denote the number of

groups that are formed for a party with n people. Thus, the expected number of groups is $E[X_n]$.

Once the first person, denoted by A, draws a name from the bag, there are two possible scenarios:

• Scenario 1: A draws her own name from the bag; this happens with probability $\frac{1}{n}$. Then, A forms a group by herself which is closed, i.e., nobody else will be able to join this group. The remaining $n-1$ people will form groups amongst themselves and the problem reduces to the case of a party with $n-1$ people, with one additional group formed already.

• Scenario 2: A draws someone else's name, denoted by B; this happens with probability $\frac{n-1}{n}$. In this case, A and B form a group that is still open, i.e., B could draw the name of another person different from A that would then join the group. Therefore, we can think of A and B as being one person for the purpose of counting the number of groups. The problem reduces to the case of a party with $n-1$ people with no group formed yet.

We obtain the following recursion formula:

$$\begin{aligned} E[X_n] &= \frac{1}{n}\,(1+E[X_{n-1}]) \;+\; \frac{n-1}{n}\,E[X_{n-1}] \\ &= E[X_{n-1}] \;+\; \frac{1}{n}, \quad \forall\, n \geq 2. \end{aligned} \tag{3.340}$$

Since $E[1] = 1$ (there is only one group formed at a party with one person), the solution to the recursion (3.340) is

$$E[X_n] \;=\; \sum_{k=1}^{n} \frac{1}{k}, \quad \forall\, n \geq 1.$$

For $n = 20$, we obtain that

$$E[X_{20}] \;=\; \sum_{k=1}^{20} \frac{1}{k} \;\approx\; 3.598.$$

We conclude that the expected number of groups at a party with 20 people is 3.6.

Solution 2: By looking at the example, we note that the groups formed, i.e.,

$$(1, 6, 10, 12, 18, 19), (2, 5, 11, 14, 20), (3),$$

$$(4, 8, 13), (7, 9, 15, 17), (16),$$

correspond to the cycles of the permutation

$$6,\ 5,\ 3,\ 8,\ 11,\ 10,\ 9,\ 13,\ 15,\ 12,\ 14,\ 18,\ 4,\ 20,\ 17,$$
$$16,\ 7,\ 19,\ 1,\ 2,$$

with each cycle written in the clockwise direction.

Thus, the expected number of groups is the expected number of cycles in a random permutation π, chosen uniformly at random from the set of all 20! permutations of $\{1, 2, \ldots, 20\}$.

In other words, if the random variable X denotes the number of cycles in the permutation π, we need to find $E[X]$.

For $1 \leq i \leq 20$, let X_i be the random variable given by $X_i(\pi) = \frac{1}{k}$ if k is the length of the cycle of π containing person i. Note that $X = \sum_{i=1}^{20} X_i$ since the total contribution from all the people in a cycle of length k is 1 and therefore

$$E[X] \ = \ \sum_{i=1}^{20} E[X_i]. \tag{3.341}$$

Note that

$$E[X_i] = \sum_{k=1}^{20} \frac{1}{k} \cdot \mathbb{P}\left(\text{person } i \text{ is in a cycle of length } k\right). \tag{3.342}$$

We compute the probability that person i is in a cycle of length k as follows: Given $1 \leq k \leq 20$, there are $\binom{19}{k-1}$ ways to choose $k-1$ people, other than person i, for the cycle of length k; $(k-1)!$ ways to permute these $k-1$ chosen people, together with person i, to form a cycle of length k; and $(20-k)!$ ways to permute the remaining $20-k$ people. Thus, the number of permutations of $\{1, 2, \ldots, 20\}$, in which person i belongs to a cycle of length k, $1 \leq k \leq 20$, is equal to

$$\binom{19}{k-1} \cdot (k-1)! \cdot (20-k)! \;=\; 19!,$$

which is independent of k and i, and we obtain that

$$\mathbb{P}\,(\text{person } i \text{ is in a cycle of length } k) \;=\; \frac{19!}{20!} = \frac{1}{20}. \tag{3.343}$$

From (3.342) and (3.343), we find that[26]

$$E[X_i] \;=\; \frac{1}{20}\sum_{k=1}^{20}\frac{1}{k} \;=\; \frac{1}{20}H_{20}, \tag{3.344}$$

where

$$H_{20} \;=\; \sum_{k=1}^{20}\frac{1}{k}$$

denotes the 20–th harmonic number.

From (3.341) and (3.344), we obtain that the expected number of cycles in a permutation (and therefore the expected number of groups that are formed at the party) is

$$E[X] \;=\; \sum_{i=1}^{20}\frac{1}{20}H_{20} \;=\; H_{20}.$$

To find a numerical value, recall that

$$H_n \;\approx\; \ln n + \gamma + \frac{1}{2n},$$

[26]Note that, given the definition of X_i, $E[X_i]$ must be independent of i.

as $n \to \infty$, where $\gamma \approx 0.5772$ is the Euler constant. Then, $H_{20} \approx 3.598$, and we conclude that the expected number of groups at this party is 3.6. □

Question 36. Let A be the sum of the digits of 2019!. Let B be the sum of the digits of A. Let C be the sum of the digits of B. Find C.

Answer: Note that $2019! < 10000^{2019} = (10^4)^{2019} = 10^{8076}$ and therefore 2019! has at most 8076 digits. Then, the number A which is the sum of the digits of 2019!, is at most $8076 \times 9 = 72684$. Thus, the number A has at most 5 digits. In turn, the number B, being sum of the digits of A, is at most $5 \times 9 = 45$ and therefore B is a 2-digit number less than 50. We conclude that the number C, being sum of the digits of B, is at most $4 + 9 = 13$.

Recall that a number is divisible by 9 if and only if the sum of its digits is divisible by 9. Since 2019! is divisible by 9, it follows that numbers A, B, and C are also divisible by 9. Since $C \leq 13$, we obtain that

$$C = 9. \quad \square$$

Question 37. Find 2019 consecutive positive integers that are not prime.

Answer: The numbers

$$2020! + 2, \ 2020! + 3, \ \ldots, \ 2020! + 2020$$

form a sequence of 2019 consecutive positive integers that are divisible by

$$2, \ 3, \ \ldots, \ 2020,$$

respectively, since

$$i \mid (2020! + i), \quad \forall \ 2 \leq i \leq 2020,$$

and therefore are not prime. □

Question 38. Exactly 4 out of 100 coins are fake. All the genuine coins weigh the same; all the fake coins, too. A fake coin is lighter than a genuine coin. How can you find at least one genuine coin using a balance scale only twice?

Answer: Divide the 100 coins arbitrarily into 3 piles, labeled A, B, and C, of 33 coins, 33 coins, and 34 coins, respectively.

In the first weighing on the balance scale, weigh pile A against pile B. There are two different cases depending on the result of the first weighing:

• *Case 1:* If pile A is lighter than pile B (the case when pile B is the lighter pile is handled similarly by symmetry), and since a fake coin is lighter than a genuine coin, the number of fake coins in the piles A and B out of the total of 4 coins can only be as follows: $(4, 0)$, $(3, 0)$, $(3, 1)$, $(2, 0)$, $(2, 1)$, or $(1, 0)$. Thus, pile B has at most one fake coin. For the second weighing on the balance scale, we select arbitrarily two coins from pile B and weigh them against each other and we find at least one genuine coin as follows:

Case 1.1: If the coins are in balance, both coins are genuine.

Case 1.2: If the coins are not in balance, the heavier coin is genuine.

• *Case 2:* If pile A weighs the same as pile B, the piles A and B contain the same number of fake coins, and therefore the number of fake coins in the piles A, B, and C can only be as follows: $(0, 0, 4)$, $(1, 1, 2)$, or $(2, 2, 0)$. For the second weighing on the balance scale, we remove an arbitrary coin r from pile A and proceed to weigh pile B together with coin r (containing 34 coins) against pile C (also containing 34 coins). We find at least one genuine coin as follows:

Case 2.1: If pile B together with coin r is heavier than pile C, then pile B together with coin r has fewer fake coins (which are lighter) than pile C. This would happen if the initial distribution of the fake coins across the piles A, B, and C was $(0,0,4)$, in which case pile A contained only genuine coins and therefore coin r was genuine, or if the initial distribution of the fake coins across the piles A, B, and C was $(1,1,2)$ and coin r which was added to pile B before the second weighing was genuine. In either case, coin r is genuine.

Case 2.2: If pile B together with coin r is lighter than pile C, then pile B together with coin r has more fake coins (which are lighter) than pile C. This could only happen if the initial distribution of the fake coins across the piles A, B, and C was $(2,2,0)$, which means that all the coins in pile C are genuine.

Case 2.3: If pile B together with coin r weighs the same as pile C, then pile B together with coin r has the same number of fake coins as pile C. This could only happen if the initial distribution of the fake coins across the piles A, B, and C was $(1,1,2)$ and coin r was fake, resulting in exactly two fake coins in pile B together with coin r and in pile C. In other words, pile A had 1 fake coin, r, and this coin was removed and placed with pile B. All the remaining coins in pile A are genuine.

Thus, in all of the cases, we find at least one genuine coin with only two weighings on a balance scale. $\square$

Question 39. Can you design a pair of 6–sided non–identical fair dice different from the standard dice with each face bearing a positive integer and having the same probability distribution for the sum as the pair of standard dice? (In other words, there must be two ways to roll a 3, six ways to roll a 7, one way to roll a 12, and so forth.)

Answer: Yes! We will use the method of generating functions to find the two . Let the non-standard dice A and B have faces $(a_1, a_2, a_3, a_4, a_5, a_6)$ and $(b_1, b_2, b_3, b_4, b_5, b_6)$, respectively, where a_i and b_i are positive integers. Let

$$a(x) = x^{a_1} + x^{a_2} + x^{a_3} + x^{a_4} + x^{a_5} + x^{a_6} \quad (3.345)$$

$$b(x) = x^{b_1} + x^{b_2} + x^{b_3} + x^{b_4} + x^{b_5} + x^{b_6} \quad (3.346)$$

be the generating functions for the number rolled on dice A and B, respectively. When the product

$$\begin{aligned} & a(x)b(x) \\ = & \left(x^{a_1} + x^{a_2} + x^{a_3} + x^{a_4} + x^{a_5} + x^{a_6}\right) \\ & \cdot \left(x^{b_1} + x^{b_2} + x^{b_3} + x^{b_4} + x^{b_5} + x^{b_6}\right) \end{aligned}$$

is expanded, the coefficient of x^n is exactly the number of ways the non–standard dice A and B can have a sum equal to n. If the sum of the non-standard dice has the same distribution as for two standard dice, the product $a(x)b(x)$ must be equal to the product of the generating functions for two standard dice, i.e.,

$$\begin{aligned} a(x)b(x) &= \left(x + x^2 + x^3 + x^4 + x^5 + x^6\right)^2 \\ &= \left[x\left(1 + x + x^2 + x^3 + x^4 + x^5\right)\right]^2 \\ &= \left[x\left(1 + x^3\right)\left(1 + x + x^2\right)\right]^2 \\ &= \left[x(1 + x)\left(1 - x + x^2\right)\left(1 + x + x^2\right)\right]^2 \end{aligned}$$

We now identify how to distribute the factors from the product above between the polynomials $a(x)$ and $b(x)$.

Since each die has positive integers on its faces, x divides both $a(x)$ and $b(x)$. Thus, $a(x)$ and $b(x)$ must each get one of the two x factors.

Note that $a(1) = b(1) = 6$ since each die has 6 faces; see also (3.345) and (3.346). Since $(1 + x)|_{x=1} = 2$, $(1 - x + x^2)|_{x=1} = 1$, $(1 + x + x^2)|_{x=1} = 3$, it follows that $a(x)$ and $b(x)$ must each get one of the two $(1 + x)$ factors and one of the two $(1 + x + x^2)$ factors.

This leaves only the two $(1 - x + x^2)$ factors to be distributed. If we distribute one of the factors $(1-x+x^2)$ to each of $a(x)$ and $b(x)$, then $a(x) = b(x)$ and the dice A and B would be identical and the same as the standard dice. We therefore give both $(1 - x + x^2)$ factors to the generating function of one of the dice, for example to die B. Then,

$$\begin{aligned}
a(x) &= x\,(1+x)\left(1+x+x^2\right) \\
&= x + 2x^2 + 2x^3 + x^4 \\
&= x + x^2 + x^2 + x^3 + x^3 + x^4; \\
b(x) &= x\,(1+x)\left(1+x+x^2\right)\left(1-x+x^2\right)^2 \\
&= x + x^3 + x^4 + x^5 + x^6 + x^8.
\end{aligned}$$

We conclude that the sum of the non-standard dice A and B with faces $(1, 2, 2, 3, 3, 4)$ and $(1, 3, 4, 5, 6, 8)$, respectively, has the same probability distribution as the sum of a pair of standard dice. Note that the dice A and B are the only non-standard dice with this property. □

Question 40. In the two–dimensional space, for every equilateral triangle ABC, we have that

$$f(A) + f(B) + f(C) = 0.$$

What can you say about the function f?

Answer: We will show that the only function with this property is the zero function.

Let O be an arbitrary point in the plane and consider a regular hexagon $ABCDEF$ with center O.

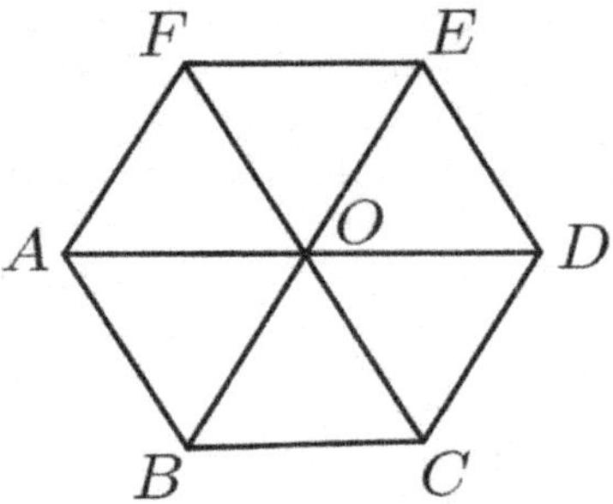

The six triangles ABO, BCO, CDO, DEO, EFO, and FAO are equilateral and therefore

$$f(A) + f(B) + f(O) = 0; \quad f(B) + f(C) + f(O) = 0;$$

$$f(C) + f(D) + f(O) = 0; \quad f(D) + f(E) + f(O) = 0;$$

$$f(E) + f(F) + f(O) = 0; \quad f(F) + f(A) + f(O) = 0;$$

By summing up all these equalities, we find that

$$2(\, f(A) + f(B) + f(C) + f(D) + f(E) + f(F)\,) + 6\, f(O) \;=\; 0. \qquad (3.347)$$

Note that the triangles ACE and BDF are also equilateral. Thus,

$$f(A) + f(C) + f(E) \;=\; 0; \qquad (3.348)$$
$$f(B) + f(D) + f(F) \;=\; 0. \qquad (3.349)$$

From (3.347), (3.348), and (3.349), it follows that

$$f(O) \;=\; 0.$$

Since $f(O) = 0$ for any point O in the two–dimensional plane, we conclude that the only function f with the required property is the zero function. □

Question 41. Suppose you have in your possession an incredibly large bag of M&M's containing a uniform distribution of the six M&M colors. (M&M's come in blue, orange, green, yellow, red, brown.) You decide to play a game: you draw one M&M from the bag and place it on the table. You then continue to draw M&M's from the bag one at a time. If you draw an M&M that is the same color as one already on the table, you eat both of them. Otherwise, you place the M&M on the table along with the others of different color. The game ends when you have six M&M's (all of different colors) on the table. How many M&M's should you expect to eat playing this game?

Answer: Denote by S_n the state with M&M's of n distinct colors on the table, for $0 \leq n \leq 6$. Let Δ_n be the expected number of M&M's that have to be eaten in order to reach state S_{n+1} from state S_n. The question asks us to find

$$\sum_{n=0}^{5} \Delta_n.$$

Note that $\Delta_0 = 0$: Starting from state S_0 with zero M&M's on the table we draw a single M&M which is, by default, of distinct color, and we end up in state S_1 with no M&M's eaten.

Suppose we are in state S_n, $1 \leq n \leq 5$. On the next draw from the bag, two different instances can occur:

- with probability $1 - \frac{n}{6}$, we draw an M&M of a color that does not match any color of the M&M's already on the table, hence moving to state S_{n+1} with no M&M's eaten in the process;
- with probability $\frac{n}{6}$, we draw an M&M of a color that matches the color of another M&M already present on the table. In this case, we eat both M&M's with matching color and we are left with $n-1$ M&M's of different colors on the table, which corresponds to state S_{n-1}. Then, in order to reach state S_{n+1} from state S_{n-1}, we will have to eat an additional $\Delta_{n-1} + \Delta_n$ M&M's, on average.

In other words,

$$\Delta_n = \left(1 - \frac{n}{6}\right) \cdot 0 + \frac{n}{6} \cdot (2 + \Delta_{n-1} + \Delta_n)$$

and therefore

$$\Delta_n = \frac{n}{6-n} \cdot (2 + \Delta_{n-1}), \quad \forall\, 1 \leq n \leq 5. \tag{3.350}$$

Since $\Delta_0 = 0$, we obtain from the recursion (3.350) that

$$\begin{aligned} \Delta_1 &= \frac{2}{5}; \quad \Delta_2 = \frac{6}{5}; \\ \Delta_3 &= \frac{16}{5}; \quad \Delta_4 = \frac{52}{5}; \\ \Delta_5 &= 62. \end{aligned}$$

We conclude that the expected number of M&M's eaten in order to reach state S_6 from state S_0 is

$$\sum_{n=0}^{5} \Delta_n = 77.2. \quad \square$$

Bibliography

[1] Avner Friedman. *Stochastic Differential Equations and Applications.* Dover Publications, Mineola, New York, 2006.

[2] Paul Glasserman. *Monte Carlo Methods in Financial Engineering.* Springer-Verlag New York, Inc., New York, 2004.

[3] Roger B. Nelson. *An Introduction to Copulas.* Springer, New York, 2006.

[4] Dan Stefanica. *A Mathematical Primer with Numerical Methods for Financial Engineering.* Financial Engineering Advanced Background Series. FE Press, New York, 2nd edition, 2011.

[5] Dan Stefanica. *Numerical Linear Algebra Methods for Financial Engineering Applications.* Financial Engineering Advanced Background Series. FE Press, New York, 2014.

[6] Paul Wilmott. *Frequently Asked Questions in Quantitative Finance.* John Wiley & Sons Ltd, Chichester, West Sussex, 2nd edition, 2009.

Made in the USA
Monee, IL
13 September 2026

62981972R00184